gasification

succeeding with small-scale systems

Andrew N. Rollinson

Published in May 2016 by

Lowimpact.org

www.lowimpact.org

ISBN 978-0-9549171-1-1

Picture acknowledgements
        Photos: Andrew Rollinson
        Fig. 19: Andrew Rollinson / Elsevier

contents

illustrations

about the author

Andrew Rollinson was born into a Yorkshire coal mining community in 1970, and the intention was for him to spend his working life down t' pit. The Thatcher government had other plans, and with its rapid dismantling of the British coal mining industry, there was no job for Andrew to go to when he left school at fifteen.

With no qualifications, there followed a series of varied occupations (butcher, labourer, gardener, policeman and more) during which Andrew stepped in and out of education until eventually obtaining, at the age of forty-one, a PhD in sustainale energy engineering. He stayed in acadæmia, specialising in the research and teaching of small-scale sustainable reactors. This led him to gasification.

The gasification world is a small one, with much interest and few specialists. Andrew began to travel Britain speaking to other gasifier owners and interested parties, sharing experiences, and knowledge. These travels also took him abroad, to Scandinavia, Germany, and India, while at the same time working on a modern gasifier system from the USA at Nottingham University. In addition to this he began assisting the UK Without Incineration Network as a technical advisor.

Andrew is a Morrisian socialist and Christian who presently combines research, teaching and writing with practising self-sufficiency at his family home. He is interested in any project that can provide local community independence, encourage sustainability, protect nature, and help others.

He can be contacted via andrew@blushfulearth.co.uk.

1. preface

"....amazing performances of gas producer-engine systems have been reported and [it is] confirmed reality that trucks have been operating on producer gas for over 300,000 km with no major repairs and less engine wear than obtained from diesel fuel. Large Italian rice mills have gasified their rice husks and used the gas to drive the power units used for milling for decades prior to World War Two. The number of quite satisfied owners of small and large gasifiers is certainly not small and there is lots of evidence that it can be done. The history of gasification has also shown that it is not one of the most convenient technologies, but in a time with less fossil fuel available and costing more each year, convenience will be a luxury that cannot be afforded very much longer."
Kaupp, A. State of the art for small-scale gas producer-engine systems, (1984).

Small biomass gasifiers were used extensively prior to and during the Second World War. Europe was prominent in gasifier development and the British government advocated the use of the technology in their colonies (1). It may then seem perplexing as to why, one hundred years later, the same country's Renewable Energy Roadmap describes gasification as "under development" (2), and that their 2012 Bioenergy Strategy, refers to it as an "advanced conversion technology" (3). Both of these documents are firmly focussed on large scale centralised power generation; but, political agendas aside, the fact is that the biomass gasification landscape with respect to performance is presently very cloudy, not just in Britain, but all over the world. Compared to other small-scale renewable technologies, gasifiers are currently much less well defined in public knowledge.

A cursory read around the subject may reveal that gasification is a crude contrivance which has little left to disclose. Look a bit deeper and a very odd picture appears of a technology that is acclaimed by some as being the answer to all future waste problems, while there are others who believe that gasification has inherent problems and should be avoided. For those considering buying, investing in, or simply trying to understand the merits of any small gasifier, the answers are very hard to find indeed. The interested party will encounter commercial information in which are extolled the virtues of some proprietary gasification system and concept, while at the same time there seems to be a complete dearth of proof. The astute reader may then perceive the question, "if gasifiers work, where are they all?" Yet, no one would argue against the well documented one hundred year history of gasifier application.

One of the problems at present is that the truth about gasifier performance and the amount of care needed to achieve successful operation is not made fully transparent by system retailers, or subsidiaries in the supply chain:

"It can not be denied that many of the difficulties [with gasifiers] are due entirely to incompetent operators. Some plants have been put out of commission temporarily by the prejudices or the lack of ability and training of the operators or engineers in charge. But, many of them have undoubtedly been the result of a short sighted policy on the part of some manufacturers, who are not willing to give proper and necessary information about design, construction, and operation of the plants made by them. The possibility of a sale at the time is apparently the only interest they keep in mind, and the future is allowed to take care of itself". (1)

This extract was actually written in 1909, following a survey of seventy gasifier plants in the U.S.A. When re-quoted in 1984, it was observed that "the situation is much the same". Today, the public perception of gasification is still a mixture of ignorance by many, but also disdain by some who have had their fingers burned (financially, not literally), optimism by others who have seen or heard about its potential but never seen or operated one, and finally a small pocket of people who have first-hand experience of successfully operating a biomass gasifier. This latter group know that a gasifier can provide sustainable energy and detachment from all the financial, supply, and environmental encumbrances associated with fossil fuels. They have, without a fanfare, succeeded with a technology that is not designed to break after a few years so that it will go to the landfill site while the manufacturer's newest model is on its way. They know however that success comes with a high input of time and effort, undoubtedly more so than they thought. But as with any project or self-sufficiency success, it has its rewards in terms of independence, pride, and the knowledge that you are living a way that is less harmful to future generations.

Consider this example from Western Australia in the 1930s. The time was the height of biomass gasifier innovation, in a part of the world where oil was in short supply. Farmers in particular were experiencing severe economic difficulties. To address this, many made use of a local surfeit of waste wood and rapidly converted their tractors to run on wood gas. There were failures, and also people who invested in the manufacture and sale of these systems without sufficient expertise. This led to many dissatisfied people who had given up after short trials. However, there were some who persevered; these being farmers who had the ingenuity and practical engineering skills to understand the system, or perhaps just the fortitude, resilience or dogged determination to make their investment work. These few people succeeded and their gasifiers gave them years of satisfactory operation (1).

I can substantiate the truth of this from my own experiences of operating small biomass gasifiers. I have seen gasifiers working 24 hours a day using waste wood cuttings, and I have evidence of modern cars being driven on 3000 km round trips with wood gasifiers providing the fuel. I have heard stories of retired experts called in to see a mis-functioning system and within an hour they had fixed it; and I have also seen many instances of where gasification has gone wrong.

Up to the 1940s, although the underlying science was poorly understood, much was achieved by diligence, ingenuity, and practical testing. Since then, some progress has been made in building up a deeper understanding but still the finer details of what goes on inside a gasifier, and more importantly how to ensure that the conditions are maintained are neither trivial nor mastered. Because of the technology's potential to theoretically turn any carbon-containing waste (not just wood, but old clothing, leather, plastic, etc) into energy, research on gasification has been steadily increasing in the last few decades. These attempts to elucidate the finer details are accompanied by the commercial world looking to get in on the ground level and make a successful investment.

But, modern experiences are hugely under-reported, expertise is hard to find, and despite there being one or two recent textbooks, descriptive quality on practical application is usually poor. Fortunately, modern systems are almost all still based on the tried and tested designs from the 1920s, 30s, and 40s, and although the pioneers in gasification are now dead, some documented records of performance along with case studies exist. This literature is spread over a hundred years, and mostly out of print, but some of it has been made available by enthusiasts, translated into English and online, or can be accessed through libraries. This literature is listed in the Bibliography.

I have not tried to improve on these old books. I have however attempted to integrate their work in a modern context for a readership that has to contend with dwindling fossil fuel reserves, increased energy prices, climate change, mountains of waste, and the modern version of a small-scale gasifier system. This book has therefore been written to provide a helping hand, drawing from recent experimental research and examples from my own practical experience on a number of commercially available gasifier systems. In one sense therefore it is focussed on educating the operator. Its aim is to empower the reader so that they can understand the technology and make an educated decision about whether they consider it right for them. Perhaps more importantly a focus of this book is that it should serve as an operational guide, so that the gasifier operator can know what can go wrong, how to avoid these problems, how to overcome them, and how to achieve long-term operational success. For the person who wants to build their own gasifier, the patents have now lapsed for tried and tested designs, and the sizing of components are freely available and provided in the Bibliography, plus Appendix A. For those intending to buy, this book will provide the necessary background information.

The definition of "small-scale" is vague and ambiguous, and I offer little to set the boundaries of constraint. I have taken the unspecified approach to designate this term to systems that are local, communal, off-grid, accepting local resources, and essentially operating with low environmental impact. Anything under 500 kW maximum rated capacity can therefore be considered as "small-scale" if it is a system supplying power and heat to a community or other large building. Predominantly however, my rationale for this book was for the 10 to 250 kW system, able to power a vehicle or a small group of homes/small farm. The size of such a system will be dictated by both the peak energy demand and also the availability of wood fuel/feedstock.

A gasifier is presently not a "push button and leave" technology. Consequently operating a gasifier will involve some continuous input of time and effort. As anyone who has had the satisfaction of installing their own wind turbine, solar panels, etc, harvesting their own crop of vegetables, or even fixing the smallest of components in their own home will know, the liberation and joy that comes with independence is the reward. It is something our ancestors knew, but which modern society has lost in no small way. A small biomass gasifier can facilitate such an independent lifestyle, and it is sincerely hoped that this book will provide the necessary understanding of a system that is in one way simple and in another very misunderstood with presently little or no support structure available.

This book will not cover industrial gasifiers such as fluidised bed systems which are electrically heated and designed to use pulverised coal, but which can use biomass. However, mention will be made of these in places to explain why small-scale systems operate in the way that they do. It will however provide a background understanding of these systems such that the reader can explore them in their own way. The focus will be the type of "gasification" system that can sustain small-scale local decentralised communal energy. This encompasses both stationary and vehicular mounted systems, as aside from a few minor differences (which will be discussed in separate sections), the scientific, engineering, and operational principles are identical. These small gasifiers are of a type called "fixed-bed" reactors, which means that the feedstock rests on a grate in a fixed position. Often they are also called "packed bed" reactors because the feedstock rests upon the grate which becomes loosely "packed" due to gravity. Neither of these descriptions are accurate, for the bed of feedstock is actually moving gradually through the reactor under gravity, and the bed must not be packed but instead it must retain spaces between the solid pieces which allow free movement of gas. The direction of gas flow is fixed, but different designs have it flowing either up, down, or across, and it is this feature that leads to the sub classification of small gasifiers.

The term "feedstock" (material that is put into a system to create a product) is often used synonymously, or at least interchangeably, with "fuel". But, as a gasifier also produces a "fuel" (the gas), for clarity and consistency in this book "feedstock" will describe only the material which is used to generate that gaseous fuel. The feedstock of a gasifier must be solid. Coal and coke can be gasified, but because this book is devoted to sustainable energy production, these will only be mentioned where relevant to explain either the history or the influences that they have on gasifier operation. Most of the time, this will mean that scraps of woody biomass are used, although other materials such as nut shells and corn cobs work as effectively if not better. In theory, gasifiers could take any solid that contains carbon, such as plastics, textile, and mixed waste; so some comment on other feedstocks is included.

Four hundred years ago, the originators of modern scientific enquiry saw no boundary between different fields of science, and neither should a gasifier operator. Gasifiers are multidisciplinary and multivariable so it is essential to be familiar with some (but by no means all) aspects of biology, physics and chemistry, as well as the exploration of practical mechanical, chemical, and electrical engineering to adequately understand and make a success of operating a gasifier. It has been necessary therefore to introduce numerous engineering and scientific terms in this book. To avoid banality for some readers, these are defined in the Glossary or in separate Text Box sections. The book however assumes no prior knowledge, and it has been written so that it can be accessible and appealing to a wide audience: the student, the renewables enthusiast, the scientist, engineer, consultant, farmer/small-holder, forester, land manager, legislator, and lay person. It aims to fill a void by providing a comprehensive reference guide to biomass gasification that is up-to date and which gives practical advice. It is also hoped that this book will be both a pleasurable and educational journey for the reader.

2. introduction

"We are only the trustees for those who come after us"
William Morris (1889)

Gasification fits within the group of energy technologies that are considered as "renewable" because they use biomass as a feedstock. This means that the energy is obtained from the naturally occurring chemical products of photosynthesis which form and are retained within plants. Because plants need to be destroyed to obtain this energy, and due to the large amounts of energy needed globally, using biomass for energy has its sceptics. These people point to deforestation, the replacement of arable land, and the effect that this has both environmentally and on local communities when biomass is cropped on the large scale (4, 5, 6, 7). In addition to all this, biomass can only be considered as sustainable in certain circumstances: if the volume of new plant growth equals removal, and if no fossil fuels were used in its culture, transportation and processing.

Increasing scale does lead to higher monetary efficiencies as upfront and operational costs can be reduced. But, if global sustainability is the main aim, then large-scale biomass combustion is not the right choice. Grid supplied electricity has gross amounts of inherent wastage. Countries such as India have transmission losses at 20%, and for western nations this figure is on average ca. 7% of their overall supply (8). In the UK, for a total electricity supply in 2014 of 359 TWh (9) these losses equate to 25 TWh annually, and are equivalent to approximately 32 Mt of CO_2 (10). This value is more than all the emissions from HGVs, buses, railways, and domestic aviation put together, or alternatively half the total emissions from all cars (11). Despite this, many countries, such as the UK choose to promote and subsidise privately owned large scale power generators to burn biomass, and claim that it is sustainable and renewable. But, notwithstanding the wastage from the electricity grid, the way that large-scale biomass combustion is managed grossly undermines its sustainable credentials.

For the UK the reality of this scenario is that it does not have enough land to grow these energy crops, and so only 1% of the biomass used in power stations is actually harvested nationally. The rest is imported, with the majority being shipped across the Atlantic from North America, or even New Zealand (see Figure 1).

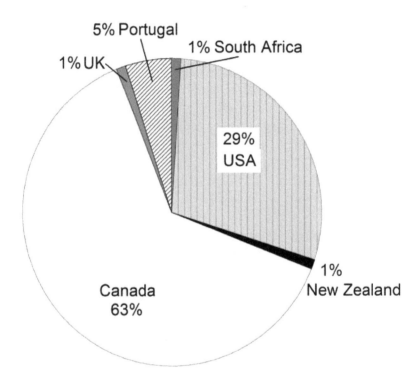

Fig. 1: wood pellet consumption for large-scale power station combustion in the UK (source data from 12, chart design adapted from 13)

The processing of this biomass is also extremely energy intensive. Large centralised power stations demand high volumes of biomass on short timescales, and from reliable sources. They prefer this fuel to be uniform, homogeneous, and easy to pulverise so that it burns quickly. Consequently almost all require their feedstocks to be pelletised (12). Due to the large volume feed rates required, temporal demands on suppliers to be reactive in response to market requirements, and the extra costs involved in locking in large areas of space, biomass supplied on the large-scale to power stations is also actively dried using ovens and large fans (14). Active dryers force air through the biomass, with usually applied heat too. The process is highly uneconomical with 3 – 4 times the amount of energy used in drying per unit gained in fuel calorific value (15). Then it is shredded, pelletised, transported, and shredded a *second* time prior to combustion.

A hierarchy of biomass sustainability has consequently been proposed (in terms of greenhouse gas savings and biodiversity goals) which puts large scale bio-crop combustion firmly at the bottom and local use of biomass waste at the top (16). It has also been stated that there is no basis for assuming that large scale use of biomass will deliver emissions savings, and in fact it is as likely that it will have the opposite effect (17).

Small-scale gasifiers are different. They make use of local biomass waste and generate energy from it at its point of use. This does not necessitate the appropriation of quality agricultural land, excessive transportation, pelletising or drying, nor the application of petrochemical fertiliser. It is actually beneficial that the biomass comes from disparate regions and in low volumes because this fits with usage requirements.

The type of biomass required for small-scale gasifiers does not necessitate the felling of trees. When land is sustainably managed or timber is harvested, anything other than heartwood is chipped because it has zero or negative value. Figure 2 shows the remains of a land clearing. The small trees were all chipped on site, but then just left in piles because landfill is now costly, where they rotted and were eventually bulldozed over. All parts of the world have this type of waste from harvesting of crops, and many countries have rural communities who burn the waste in-situ which then creates air pollution (18). Furthermore, it is environmentally damaging to do nothing with this bio-waste as an estimated 590-800 million tonnes of methane is released annually from anaerobic decomposition. Methane is a 21 times more potent as a greenhouse gas than carbon dioxide (19).

Fig. 2: wood chip left in-situ where trees have been felled

Such biomass would not be deemed exploitable for large scale supply due to the variation in its production, disparate nature, and associated economics of collection and transport. But this makes it most suitable for small gasifiers. Because woody biomass is relatively bulky to transport and store (hence the reason why large power stations shred it and pelletise it then shred it again) the beneficiaries are those closest to the resource. In rural areas this coincides with areas that are more likely to be off-grid, but which have a need for lighting and day to day activities which require electricity, motive power and perhaps also heat that a small-scale gasification system can provide.

The resource of waste biomass is significant too. In the UK, there is an annual 396,000 tonnes oven dried weight of forestry arisings produced, with an estimated 68% that does not have a market and goes to landfill (20). For woodlands, it is estimated that sustainable management can produce 3.75 oven dried tonnes per hectare per year (21), and there are a further 1.2 million hectares of unmanaged forest reportedly available for sustainable wood supply (16). A more recent study estimated 3.3 million tonnes (Mt) annually from parks and gardens much of which will likely be woody shrubs suitable for small-scale gasification (22). Then there are the high volumes of biomass produced locally as a necessary consequence of the sustainable preservation of habitats, which are reportedly the second largest sustainable biomass resource available in the UK (16). In addition to this, there is waste wood (such as from households, furniture and construction industries, and virgin wood processing), the total volume of which is estimated at a further 4.1 Mt (22, 23).

Efficiency and sustainability benefits therefore come from the close-coupled nature of biomass to conversion and energy use. Biomass gasifiers can be grid-tied, but this is unnecessary as they are able to convert biomass to power instantaneously on demand. Without the market constraints of large volumes over short lead times that large-scale generators have, gasifier owners can simply plan from season to season using locally harvested resources. And, for both rural and urban situations, it can be envisaged that the small gasifier has community enterprise potential making use of biomass waste at the local level. Based on the approximate resource values, 5.25 million tonnes would feed 152,000 small gasifiers of the type described in this book rated at 10 kW_e for eight hours per day for 360 days per year, assuming feedstock is consumed at 12 kg per hour).

3. what is a gasifier?

Perhaps five thousand years ago, the phenomenon of pyrolysis first became known. Pyrolysis is a process of heating a carbon-based substance in the absence of oxygen. It was applied to the ancient method of producing charcoal – a material useful because of its clean burning characteristics (24). In this method, a wood pile was created, then almost, but not entirely, covered with soil. The covered log pile was then ignited. The soil stopped enough air getting in to create flames, but permitted just enough for some portion of the wood to smoulder and emanate heat. Over a number of days this heat promoted the release of wood (or to be precise "pyrolysis") gas, leaving behind charcoal.

Charcoal is the structural solid "fixed" carbon framework of wood. Without oxygen being available, it is resistant to thermal decomposition and it keeps the same size and shape as the initial wood piece from which it derived, although about 80% lighter (See Text Box 1). With ancient charcoal production methods, which are by the way still practiced today, the gas is considered a waste product and just vented to the atmosphere. With gasification, the same process of pyrolysis occurs, but it is the gas which is desired. "Gasification" is not however synonymous with "pyrolysis". Pyrolysis occurs as part of the gasification process, so it is but just one in a number of chemical reactions inside a gasifier.

On its own, a gasifier is of no practical use. Its primary purpose is to optimise this thermal conversion of solid biomass into a combustible gaseous product. The main benefit of doing this is that it provides greater flexibility. The gas can be piped to a boiler and burned to produce heat, or to a stationary or vehicular internal combustion engine for conversion to mechanical power. If this engine is connected to a generator, then renewable electricity can be produced independent of the grid. Heat is an engine by-product, so the gasification system can also give combined heat and power. No external supply of oil, gas, or electric is needed as the system is self-sustaining.

In many ways a gasifier is a very simple machine. All that it requires is a container able to withstand high temperatures, such as a steel cylinder (see 25). Within this container, there occurs the "process for making gas" which involves encouraging certain naturally-occurring chemical reactions. This definition of the word "gasification" is, I feel, better than other terms which the reader may encounter such as "partial oxidation" and "sub-stoichiometric combustion". But in many ways gasification is also difficult. Ancient charcoal production methods require constant babysitting. If too much air is allowed in, then the pile can ignite, while too little and the process can stop. Between these extremes, the process can drift away from optimisation and slight changes can affect the quality of product. This delicate balance of air intake and also size of the wood pile still remains true of modern gasifier design, as will be seen.

Text Box 1

Release of pyrolysis gas is evident by the flickering flames experienced when enjoying the pleasure of an open log fire. These flames are created by the continuous burning of gas which has come out of the wood, rather than burning of the wood itself. This release of pyrolysis gas begins at about 220°C to 250°C, and being combustible it ignites when encountering oxygen in the air[1] as long as there is an ignition source such as the striking of a match or heat from the embers of a fire on which subsequent wood is placed.

---

[1] Heat must first be applied to the wood to cause the ignition and to overcome the activation energy demands of drying and pyrolysis. Thereafter, combustion is an exothermic process and it self-sustains the fire/flames (see Text Box 3).

Apart from minor quantities of inorganic elements necessary for plant growth such as phosphorus, potassium, magnesium, etc, wood is predominantly composed of carbon (C), oxygen (O), and hydrogen (H). These three main elements are taken in by the plant from its surroundings, then changed via photosynthesis into long chain organic molecules: lignin, cellulose, and hemicellulose - which are the structural components of biomass. When wood is heated, these structural components break down into smaller complex hydrocarbons and become gaseous whereupon they evolve from the carbon framework of the wood.

Phase (solid, liquid, and gas) is a function of temperature and pressure. For most substances, increasing temperature causes the phase change along the solid to gas scale, and decreasing temperature causes the opposite. The size of the molecule also determines its "state". Larger hydrocarbon molecules usually have a greater propensity to remain in solid or liquid phase, whereas smaller hydrocarbon molecules are more likely to become gaseous as temperature increases.

Scientific tests have been done to determine the quantity of volatile hydrocarbon content in many different types of biomass. For woody biomass with an average moisture content of 5-10%, subjected to slow heating without oxygen at temperatures up to 500°C these values are ubiquitously in the 70-80% weight range. What is left behind (the remaining ca. 15%) is charcoal – the fixed carbon framework of the wood along with inorganic minerals (Figure 3). The mineral content varies by species but is about ca. 1% for wood. Silica is found to be high in grassy species, and nut shells contain higher carbon and ash content and therefore less volatile matter. Slow heating optimises gas production.

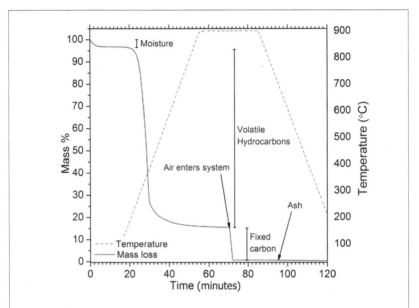

Fig. 3: thermogravimetric analysis results of a woody biomass sample being heated without oxygen. Air is introduced at 70 minutes whereupon the char is burnt leaving behind the ash component.

To maintain the gasifier reactions in steady state and to make the gas of a high purity needs the components shown in Figure 4, so four mechanical sub-systems. The reactor is where the gas is produced, so in this sense it is the "gasifier", but as the other components assist, "gasifier" is sometimes also used to describe the whole system.

To operate it, the reactor is first filled with small, chopped pieces of wood. The engine provides the driving suction that pulls gas through from the reactor, at the same time as pulling a small amount of air into the reactor core (the reasons for this will be explained in the next section). Temperature is provided by chemical reaction (again, explained in the next section). The flare is only used at startup and shutdown, and acts as a bypass for these short periods when the reactor is not at operating temperature because at these times the gas is not pure enough for the engine.

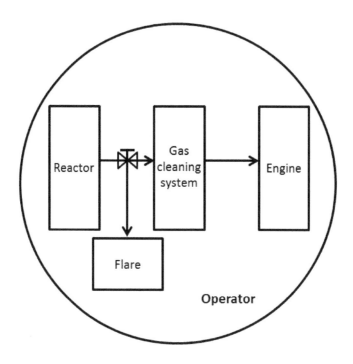

Fig. 4: simple gasification flow diagram

The small scale gasifier is built to last, from basic, and easy to change components. It has a benefit over wind and solar renewables technologies because it provides power and heat on demand rather than being limited by seasonal or daily weather variations. Small biomass gasifiers therefore don't need battery banks to store the power that they generate, and consequently there are efficiency, cost, space, and labour savings through this. They operate like a car with a fuel tank of wood chips, in which the energy is stored and used when required. With some level of automated control, the gasifier system can be turned on and the engine will idle away using wood chips at a slow rate in the morning, powering lights, fans, pumps, etc, then in the evening as a shower or a cooker is used, the generator will tell the engine that more power is needed and the engine will speed up accordingly. This will cause the wood chips to be consumed more quickly. Turn off the cooker or shower and the engine will slow down again, and the wood chips in the fuel tank will be preserved for the next switch on. For those who already have their own wind turbine/solar panel and battery backup system, the gasifier is an alternative to a diesel generator.

It should be noted at the outset that running a gasifier either stationary or vehicular will be somewhat like caring for a classic car: more intrusive on time than an electronic model, but in addition to the pleasure that comes from any self-sufficient DIY project the gasifier is a system that is serviceable "in the field" (Figure 5).

Fig. 5: re-connecting a thermocouple inside the reactor of a gasifier

Gasifiers have efficiency benefits over biomass steam turbines for electricity production, as there is no need to use water as a "working fluid". Higher efficiencies can also be achieved by being able to combust the gas at higher temperatures in engines (according to Carnot's theorem). However, this is offset by lower power outputs due to aspects related to the lower heating value of gas and air mixtures (26). This is discussed more fully in Section Engines and Generators.

To achieve gasifier success, there are three primary aspects to get right: 1. the reactor core, 2. the feedstock, and 3. the operator. The "gas cleaning system" is not one of these three and many people waste extensive resources of time, effort and money on this ancillary system in vain. Get any of these three components wrong and the gas cleaning system will never function, and subsequently the whole system is likely doomed, as gasifier history has shown.

4. history and origins of modern gasification

The history of small-scale gasification is a rich one[1]. It tells the story of a technology that was discovered, explored, and implemented, but then abruptly abandoned because something cheaper and easier to use came along. Biomass gasification soon re-appeared, but it was never wholly embraced. In cycles, interest has waxed and waned, whenever the threat of oil scarcity has occurred. Yet, throughout these periods there have been a few individuals who wanted to go their own way, and so maintained a successful interest in biomass gasifiers.

A patent from 1819 describes a gas producer system coupled to a gas suction engine. The "suction gas" description soon fell out of use to be replaced by the term "producer gas", likely coming from the fact that as soon as the gas is produced it is used. This term still applies today.

The first record of a passenger vehicle running on producer gas was in Scotland, built by J.W. Parker between 1901 and 1905. Bernier created a "suction gas producer" in 1905 which could be made small and compact, but his patent was inadequately protected and as a consequence this stimulated great entrepreneurial designs for gasifier technology. It is on these designs that the present systems are based.

Motivated by the prospect of war and consequent restrictions in oil supply, Britain put resources into vehicular gasifier design during the first two decades of the 20th Century. After the First World War, the British Government considered

---

1 The information in this section has been obtained from the following references, unless otherwise stated: (1, 27, 28)

that gasifiers were better for their colonies, where charcoal could be easily made and used as a feedstock, thereby preserving fossil fuel supplies. Germany also put emphasis on small gasifier development in the 1930s, having most success with a design by the Frenchman Jacques Imbert from the 1920s (29). This Imbert design has stood the test of time, being used in 74% of the Swedish systems deployed during the Second World War and continues to be used in most modern gasifiers.

A word on coal gasification. This was practiced initially - and remains today a method of producing high-value carbon - but coal is unsustainable and for many other reasons it is not a perfect feedstock to use in a small gasifier. It is very slow to ignite, has a higher ash and sulphur content, and it is not always available locally. Gasifiers designed to accept wood scraps and agricultural residues were therefore the obvious progression. Although wood has a lower energy density, the small quantities required for a gasifier are in most places plentiful and readily available, being close to the point where power is required, obviously not requiring any mining or delivery costs, and soon replenished by Nature. This use of biomass on the small-scale with waste wood is just as attractive today.

Between the coal and the biomass gasifiers there were others designed to use charcoal. The Swedes began introducing these in the 1920s on farm tractors, but they were also used along with wood in other parts of mainland Europe and in America at the same time. The advantage of charcoal is that it overcomes the problem of tar in the producer gas which occurs when biomass is used as a feedstock. Overcoming the tar issue is however offset by the laborious practicalities of creating the charcoal first, along with much lower overall efficiencies since about 70% of the gas producing molecules are removed during the production process (see Text Box 1). This led to the realisation that with such low efficiencies, very soon all the forests would be gone. So, during the early 20th Century wisdom prevailed and charcoal gasifier construction was prohibited in France and Denmark, and restricted in

Germany. As an alternative, the gasification of coal-based feedstocks and peat was encouraged. Problems soon ensued because of the high sulphur content of these feedstocks, along with variations in size and shape, and expensive production methods. The Scandinavian countries continued with both wood and charcoal gasification since they had greater reserves and less oil. In doing so they developed a greater breadth of expertise.

As it is relatively easy to convert a compression or spark ignition engine to run on wood gas (see Engines and Generators, and 25, 26, 27, 30), when the Second World War started in 1939 many petrol and diesel engined vehicles were converted. Records evidence approximately 1 million civilian and military vehicles (mainly charcoal), including boats, trucks, tractors (mainly wood), and cars being powered by mass produced wood gas systems (26, 31), mostly in countries where fuel shortages were severe. The Swedish parliament legislated for gasification and instigated operator training and licensing, along with certification for system manufacturers which increased consumer confidence. Germany also reduced the number of manufacturers and focussed on making only models that were proven as successful, along with setting up a national feedstock supply infrastructure through garages and filling stations.

However, after the Second World War, when oil became cheap, wood gasification use declined to practically zero from 1950 onwards except for very small pockets of interest in widely dispersed places. During these stable times, society became reliant on oil for its motive power.

Sweden was one of the few countries to continue investigations on small-scale wood gasification post 1945. At their National Machinery Testing Institute from 1957 and into the 1960s, they commenced extended tests to determine and develop a standard type of wood gasifier that could be used for vehicles and at a range of sizes (26).

In the early 1970s, the Oil Crisis provided an incentive for renewed interest in gasification, and this stimulated efforts to revisit and document the information from the World War Two period. Research continued through the 1980s with some momentum developed by a few countries which saw a benefit in the technology for security in case of oil shortages and in general off-grid power from local waste wood (25, 26, 31, 32). However, it was thought by others in 1979 that the return of gasifier to engine systems might never be needed because of ideas about electric cars, and synthetic fuels such as methanol or pure hydrogen. When these novel concepts did not progress as expected, and when global warming became apparent, the sustainable potential of wood gasification took on a new and larger significance. Research and investment in the technology has been slowly growing since then. The Scandinavians, Germans, and Dutch, along with countries in Asia, North and South America, have all been prominent in biomass gasifier research and development.

4.1. modern small-scale gasifiers

Two modern small-scale biomass gasifiers are now introduced. These will be used as exemplars throughout this book.

In 1987, the Indian Ankur Scientific Energy Technologies initiative (33) began attempts at commercialising small biomass gasification, and the reactor shown in Figure 6 is one of their systems. Today they manufacture systems of varying sizes and for various applications, supplying predominantly to South and Southeast Asia. Countries without a mature electricity grid infrastructure have an interest in gasification because it can ensure stable rural power for agriculture and industry, using locally available waste. The Ankur gasifier in Figure 6 stands ca. 2 m high and has a semi-open topped hopper which is filled with chopped wood pieces. Off picture is an engine able to take the directly supplied producer gas (usually dual fuel compression engines are used) and this engine creates a negative pressure inside the reactor with each piston stroke. This suction pulls air inward through two mid-section air holes (as shown on Figure 6) which drop down into the reactor core in a narrow central region. This design is for chopped wood, but Ankur also manufacture a different configuration for smaller feedstocks, e.g. rice husks.

Fig. 6: Ankur downdraft gasifier showing two air inlets which take the air to the central section of the reactor

The second example is more recent, and is made by GEK-All Power Labs (system shown in Figure 7). This Californian company initially ran an art studio and built a wood gasifier when their electricity was cut off by the local authority. From this they began making and selling Gasifier Experimenter Kits (GEK) for universities, which was followed by expansion into the production of small-scale stationary gasifiers of varying capacity and complexity, initially for enthusiasts and farmers (34). Their system is also a fixed-bed and "throated" downdraft type based on a design from the 1920s and 30s, but has a number of heat exchangers and proprietary electronics which monitor system variables so that the reactor performance can be observed and maintained to some extent.

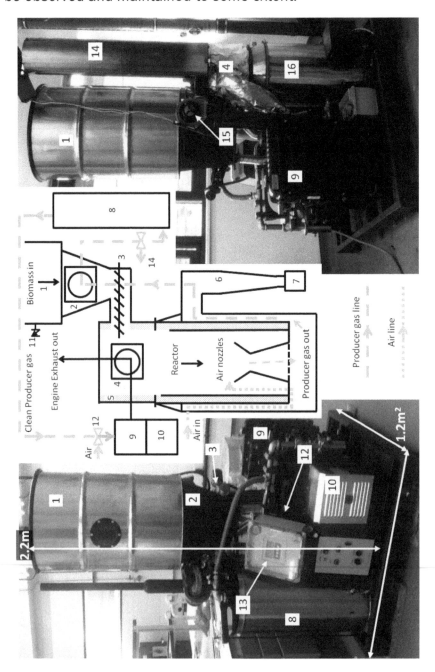

Fig. 7: see next page for key

Fig. 7: the Power Pallet, Mk 4. Schematic is not to scale 1, feedstock hopper; 2, auger channel heat exchanger; 3, auger feeder; 4, heat exchanger; 5, insulation; 6, cyclone; 7, cyclone particle collector; 8, gas filter; 9, three cylinder gas engine; 10, 10 kWe generator; 11, pressure release valve; 12, engine intake air:fuel valve; 13, electronic control system; 14, flare; 15, fan blower; 16, reactor

5. principles of gasification

"Gasifiers are relatively simple devices. The mechanics of their operation, such as feeding and gas clean up, also are simple. The successful operation of gasifiers however, is not so simple. No neat rules exist because the thermodynamics of gasifier operation are not well understood. Non-trivial thermodynamic principles dictate the temperature, air supply, and other operating variables of the reactors...it is a tribute to the persistence of experimentalists that so much progress has been made in the face of so little understanding." (31)

The information in this section is background science. It is relevant to all the subsequent sections and provides the operator with knowledge for successful gasifier operation. It can of course be referred back to and it has been structured with this in mind.

Thermal decomposition of any material is a very complex mechanism involving hundreds of chemical reactions and producing a mixture of solid (*e.g.* soot), condensable ("bio-oil" or "tar"), and gaseous constituents in various combinations. These all originate from the initial pyrolysis process.

An optimised biomass gasifier has an ideal gas composition of carbon monoxide (CO) and hydrogen (H_2) at ca. 20% each, with lesser concentrations of carbon dioxide (CO_2) at ca. 5 to 10%, some methane (CH_4) less than 3%, water vapour (H_2O), and oxygen (O_2) less than 2%. Apart from other trace molecules (less than 1%), the remainder will be nitrogen (N_2) which comes from the air used in the process (air being 79% N_2). The N_2 is inert, so non-combustible, and its influence on the process is that it merely dilutes the gas. It also passes through the engine or combustion chamber unchanged. Water vapour will mostly be removed prior to the engine too, leaving a combustible mixture of predominantly CO and H_2 diluted by N_2. Gas mixtures with this sort of composition are termed syngas (or synthesis gas) which refers to their application in chemical processing. Although the name syngas is often used to describe gas from a gasifier, this is only because of similarity in composition. To be precise about this, syngas is the _type_ of gas that is created by a gasifier, but producer gas is the name only for the gas which _has been made_ in a gasifier.

The contaminant molecules within the gas, soot and tar are between 5 – 20% for an updraft gasifier (50,000 – 200,000 ppm) and 0.01 – 0.1% for a downdraft gasifier (100 – 1000 ppm) (31). Both types of impurity are detrimental to long term operation of an engine, and unless the producer gas is sent directly to a burner, these molecules will therefore need to be cleaned out. Even a well designed gasifier using good feedstock will still need a gas cleaning system. But this sub-system will only function adequately if the reactor core is operating properly, and if it has an operator who is both diligent and able.

So, the chemical aim of the small biomass gasifiers described henceforth in this book is to encourage the certain favoured reactions, and to produce, at steady state, a gas which is rich in H_2 and CO, plus (if sent to an engine) depleted as much as possible in tar and soot. By going into details about small-scale gasifiers, it is necessary to introduce the fact that these gasifiers are designed in part, with tar reduction in mind, and to have one operating successfully will mean focussing on how this is achieved in the gasifier core. Tar and soot will become an increasingly common subject as this book progresses, and Controlling Tar and Soot Formation is devoted to this specifically.

5.1. gasification reactor thermochemistry

It helps to consider a gasifier reactor as being separated into zones. These zones are not physical compartments, as there are no internal segregations (see

Figure 8 and Figure 10). The zoning is created by regions of different temperature, achieved by controlling the oxygen intake because oxygen, through combustion, generates regions where heat is produced (see Text Box 2).

In each zone, different temperature-dependent chemical reactions occur and need to be nurtured. Reactor sizing in relation to the size of the air intake holes is crucial, as the amount and location of incoming air dictates the size, position, and extent of temperature fields which define each internal zone. This need not be a concern, as, from numerous experiments in the early part of last century, the ideal sizings of gasifier reactors have been determined and these are provided in Appendix A.

Fig. 8: looking from above into a gasifier reactor, the narrow "throat" section is discussed in The Downdraft Gasifier . The tongue-shaped protrubrance from the top middle is the auger paddle switch from the system shown in Figure 7. Ignore the blue fluffy deposits which came from my failed attempt to gasify recycled textile

To illustrate the principle of thermal zoning, Figure 9 shows the simplest type of gasifier: the "updraft". Its name comes from the orientation of air and gas flow, from bottom to top, and importantly, countercurrent to the gravitational throughflow of biomass. Once inside the reactor and exposed to heat, the biomass thermally decomposes until all that is left is ash, which falls through the supporting grate at the bottom. If all this sounds somewhat familiar it is because an everyday form of updraft gasifier is a cigarette. It is filled with biomass and ignited at one end (the combustion zone in Figure 9). Once ignited, the smoker draws air from the opposite end to pull oxygen into the combustion zone to maintain the process and also to pull the tobacco gases out.

Text Box 2

In schools, even at an advanced level, chemistry is taught by elemental reaction scheme notation. This is an oversimplification. The chemical reactions that constitute combustion, or any but the most simple process (and combustion and gasification are not simple processes), involve many reactions and half reactions which comprise transient chemical species. It need not concern the reader as the general reactions given here will suffice. If the reader is not familiar with this, or with chemical symbols, molecules and elements, explanation is given here and in the Glossary.

R1, for example is a simplification of the overall chemical reaction that occurs when carbon (in charcoal) burns in air (with oxygen).

$$C + O_2 \rightarrow CO_2 \qquad \Delta H = -394 \text{ KJ/mol} \qquad R1$$

To the right of the line on which R1 appears. The triangle symbol is the Greek letter delta. It denotes "change". When combined with the capital italicised H ("enthalpy"), these two symbols describe heat change at constant pressure.

The numerical value quantifies this change in enthalpy when the reaction proceeds in the direction of the arrow shown between the elemental and molecular symbols on the left hand side. The greater the negative value here, the more energy that is released by the reaction. Similarly, the greater the positive value, the more energy that is *absorbed* by the reaction.

A chemical reaction that gives off heat is "exothermic", from the Greek "exo = external", *therme* = heat. Combustion is such a chemical reaction. Reactions that absorb heat are "endothermic", "*endo* = within". Making water boil is an endothermic reaction. Burning coal or wood is an exothermic reaction.

So, in the case of R1, one atom of carbon reacts with one molecule of oxygen to form one molecule of carbon dioxide and at the same time releasing 394 kJ of energy. It gets more complicated than this (in terms of energy release) as the temperature at which the reaction occurs alters the energy emitted or absorbed, but that is not necessary for this book.

The relevance of all this is to show which molecules react and to compare the different changes of energy needed for each of the different type of reaction. This is to help understand how gasifier reactions create a balance of internal energy and also how to identify, avoid, or solve problems, many of which are due to temperatures and therefore non-optimisation of chemical reaction conditions.

With the updraft gasifier, air is allowed to enter at the base, and at the top there is some mechanism of active suction to pull the producer gas through. In other types of gasifier, an engine can serve the purpose of generating the suction to drive the reactor because of the natural vacuum created by its reciprocating pistons. The engine has a dual purpose of also utilising the gas which is produced. But the gas from an updraft gasifier is too dirty to be sent to an engine, the reasons for which will be explained in greater detail later.

At the top of the updraft gasifier (in Figure 9) is biomass that is furthest away from the hot region and so least chemically disturbed by temperature. As this biomass gradually moves closer to the combustion zone it is subjected to an increasing temperature gradient whereupon it loses first its moisture, then its volatile component, and finally the char is burnt, leaving just ash. High combustion zone temperatures will ultimately burn the carbon content of the char to heat the rest of the gasifier until these char pieces get so small that they fall out through the grate.

It is important to stress that the exothermic nature of combustion provides all the heat energy for the other zones which are all "endothermic". Consequently, and by design, the whole reactor is maintained in a state of thermodynamic equilibrium merely by optimising the size of the air inlet and air flow in relation to reactor size. This is how it operates itself without any external energy. When a chemical reactor is able to do this, the process is said to be "autothermal".

Updraft gasifiers are thermally efficient, but they are chemically inefficient. To understand why, consider that there are a finite number of H, C, and O elements available in the wood. The gasifier operator wants as many of these elements as possible to form CO, and H_2 (but if some form CH_4, then it's okay). Because of this finite amount of C,H, and O available, if some form tars, then that will mean that there must be less CO, and H_2. Updraft gasifiers also have problems with ash clogging the grate as the ash can melt at the very high temperatures produced when char burns. The updraft gasifer is however a simple design and it provides a good way of illustrating the different zones.

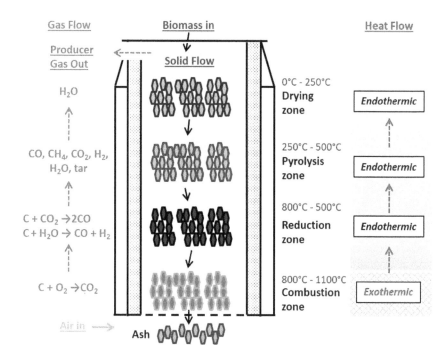

Fig. 9: updraft gasifier, showing the thermochemical flows

5.1.1. combustion zone

To understand gasification, it is important to know the details of combustion. The lit end of a cigarette (or combustion zone of an updraft gasifier) does not burn with a flame because the pyrolysis gases are being pulled back in the opposite direction to the incandescent hot section that is open to the oxygen in air. What is burning at the lit end of a cigarette is the char (specifically the fixed carbon) of the biomass, by reaction R1. All the original volatiles (water and pyrolysis gases) have been removed in earlier zones and only this pure carbon (and the mineral ash) is left behind. In an open log fire, it is the burning of only this fixed carbon that occurs at the end stages, when the fire is down to its embers, glowing red hot but without any flame. Here the charcoal is burning hot and pure because all the pyrolysis gas has evolved, hence charcoal is created for metallurgical smelting and for cooking.

$$C + O_2 \rightarrow CO_2 \qquad\qquad \Delta H = \text{-394 KJ/mol} \qquad R1$$

The combustion reactions have an energy barrier (called "activation energy"), which needs to be overcome first (see Text Box 3) and the amount of activation energy can be determined by experimentation. It has been found to vary with each type of reaction and each type of wood species being burned (or gasified). It the wood is wet, then the ignition source will not overcome the activation energy needed for the gas to ignite, which is why wet fires smoke. In these instances the wood is being heated, and releasing its gas, but the gas (being hot and buoyant) rises upwards accompanied by particles of soot (the smoke). But, if the conditions are right, such as if the wood is dry enough, and if there is sufficient air circulation, chemical reactions such as R1, R2, and R4 will become self sustaining, and the fire is then "going".

Text Box 3

For a chemical reaction to occur there often needs to be an input of energy (called the activation energy). It is an energy barrier that needs to be broken, an example of which is applying a match to some paper to set it alight, or igniting the natural gas on a cooking stove. This applied heat breaks the activation energy barrier, so that the reaction can occur. Some reactions then give off heat (exothermal) so that further reactions propagate.

Once the activation energy barrier has been overcome all reactions could potentially move both ways, although chemical reaction nomenclature often shows only uni-directional arrows for simplicity and in these cases the forward (or reverse) direction is overwhelmingly favoured. Some chemical reactions are easily reversible, while some are essentially irreversible. The reason for this is due to something called "entropy" which is outside the scope of this book.

Later you will encounter a reversible reaction called the "water gas shift" (R9). This reaction is reversible and when moving forwards it is exothermic (so that it gives off heat when it moves to the right).

$$CO + H_2O \leftrightarrow CO_2 + H_2 \text{ (water gas shift reaction)}$$
$$\Delta H = -41 \text{ KJ/mol} \qquad R9$$

A chemical reaction will stabilise and can be at "equilibrium" at any temperature, but temperature (and pressure, and whether the products and reactants are removed) dictate whether a reaction is promoted to the right – forwards (so more CO_2 and H_2) or to the left – backwards (more CO and H_2O).

By Le Châtelier's principle, which applies to the rate of change in chemical reactions, endothermic reactions are pushed to the right by higher temperatures, but exothermic reactions are pushed to the left. The water gas shift reaction therefore begins to favour its reverse direction at higher temperatures (above say 200°C). So, if this reaction can be made to reach equilibrium far to the right it will create more H_2 product. But the higher temperatures work against this and the only way that at 900°C this reaction could be made to move significantly to the right is if both the products were instantaneously removed as fast as they were created, again by another aspect of Le Châtelier's principle.

A defining aspect of a gasifier is that _only limited amounts of oxygen_ are allowed to enter the reactor. Otherwise the system would operate more as a combustion chamber/incinerator, burning rather than gasifying the wood and pyrolysis gases inside. This would then lower the yield of combustible gas that is produced. With some larger industrial-scale (fluidised bed or entrained flow) gasifiers which are electrically heated, the amount of air or pure oxygen (some even use steam) which is allowed to enter is actively introduced and its flow rate can be controlled to adjust the desired concentrations. For the small-scale fixed-bed gasifier, the air cannot be controlled in this way, and optimisation of air intake is achieved through design, based on the strength of suction, size of reactor, and cross-sectional area of the air inlet.

Tests have shown that the best conditions for gasification are achieved at about 24% of the oxygen that would be required for full combustion of the biomass (31). The amount necessary for full combustion is called the "stoichiometric" amount and is based on the chemical components of the biomass that would otherwise be fully combusted via R1 and also R2. Gasifiers therefore operate with sub-stoichiometric air, and, the ratio between actual air used and that which would be required for full combustion is known as the gasifier "equivalence ratio".

$$H_2 + \tfrac{1}{2}O_2 \rightarrow H_2O \qquad \Delta H = -242 \text{ KJ/mol} \qquad R2$$

5.1.2. drying zone

In a gasifier, a log fire, match, cigarette, or any carbon-based material to which heat is continuously applied, the first stage of thermal decomposition will cause internal moisture to heat up and change phase to a gas (R3), whereupon it evolves. This occurs in the drying zone of a gasifier (Figure 9).

Wood inherently contains moisture even if it may appear dry on the surface. It is the loss of this cellular moisture which is achieved by slow outdoor seasoning. The reason that moisture dampens out combustion is because the heating up of H_2O absorbs lots of energy, partly because of reaction R3 when it changes phase from water to steam. Text Box 4 explains why the moisture content of any feedstock either in gasification or combustion is described as imposing a "parasitic heat load". This has a major influence on operational performance because of the impact that it has on adjusting internal temperature; and the quality of gas from a gasifier is more dependent on temperature than any other factor (35).

$$H_2O \text{ (liquid)} \rightarrow H_2O \text{ (gas)} \qquad \Delta H = +41 \text{ KJ/mol} \qquad R3$$

In the updraft gasifier (and in a cigarette) this zone is the last in sequence prior to the gas exit point. Consequently since heat has to travel through the other zones (both of which also involve endothermic reactions), the temperature here is lowest, but still above 100°C, and still below $220 \leq °C \leq 250$ when hydrocarbons begin to evolve. This higher temperature marks the boundary between drying and the pyrolysis zones.

Water has extremely unusual properties (36). These mostly come from the bonding between hydrogen and oxygen. It is general knowledge that water boils and becomes steam at 100°C. However, if an open glass of water is left in a room, within a week or so, the water will have disappeared. It will have "evolved" into the atmosphere, even though the temperature didn't get anywhere near to 100°C. So, when I say that moisture in the wood is released at ca. 100°C, this means that above 100°C, *most* of the water will rapidly evolve.

The updraft gasifier (and the cigarette) design is a thermally efficient configuration because the parasitic heat load from internal moisture does not affect the other zones. H_2O passes out through the exit instead of passing through the hot zones where it would otherwise absorb energy. So, it is pertinent to explain why updraft designs are considered a *less* attractive form of small gasifier. The answer lies in the location of the pyrolysis zone (Figure 9), which is, as with the drying zone, downstream of the high temperature regions. All the pyrolysis gases, that in an open fire are burnt within the combustion flame, must consequently come out in their raw state. Because of this, using an updraft gasifier to supply producer gas to an engine is chemically _inefficient_, since ca. 70% of the wood mass which is released as a volatile gas comes out "unpurified". As an aside, it is the carcinogenic nature of these tars which cause lung cancer for smokers.

5.1.3. pyrolysis zone

In a gasifier pyrolysis zone, the biomass has lost its moisture and is further along the thermal gradient created by the exothermic combustion zone. Being nearer and therefore hotter, the biomass encounters these higher temperatures until it reaches a stage where volatile hydrocarbons are released. Thus, begins the evolution of pyrolysis gas: CO, CH_4, CO_2, H_2, H_2O and tar vapour by R4. Given sufficient time and a constant source of temperature pyrolysis becomes complete by ca. 500°C (although minor quantities of pyrolysis gases are still evolving up to 800°C (37)), leaving behind just solid charcoal. Reaction R4 simplifies this phenomenon, known as "pyrolysis" by way of a global irreversible reaction where heat is causing the biomass (represented by $C_6H_{12}O_6$) to decompose. R4 is an endothermic reaction, thus absorbing heat from the neighbouring higher temperature zones. No value is given in R4 because the reaction scheme shown is greatly simplified. Pyrolysis gas is a mixture of many kinds of molecule, most of which are long polymers and the reactions that form and re-form them are complex (see Controlling Tar and Soot Formation). This zone is therefore the gas producing section of the reactor. In a gasifier the pyrolysis gases are not fully combusted, as the restricted oxygen has not enabled a flame. So the gas is rich in the partial or un-oxygenated products from thermal decomposition.

$(C_6H_{12}O_6)_x \longrightarrow$ heat without oxygen
$(H_2O + H_2 + CO + CH_4 +.. C_5H_{12})$ $\qquad \Delta H = + -ve \qquad$ R4

Pyrolysis gas is always created and released from biomass when it is heated above ca. 250°C. Even in an outdoor fire, which has practically infinite quantities of oxygen to draw upon, pyrolysis occurs. Pyrolysis gases are visible when the fire is being started, since before the gas has ignited it is evident in the form of smoke and soot (which are both condensed hydrocarbon particles), and steam (from R2) rising buoyantly into the air. Once ignited, these initial stages of pyrolysis go unseen because the gas is being produced more rapidly than the oxygen can reach it, so that the gas exists but only for only a fraction of a second before being burned.

The combustion potential of a fuel is usually described as its "calorific value", which is calculated from the sum total amount of energy released from all the component molecules (ΔH =+ve), minus any water (ΔH =-ve). A high purity producer gas is therefore one that is rich in molecules such as CO (which can be made to release energy via (R5), H_2 (releases energy via R2), and CH_4 (R6).

$$CO + \tfrac{1}{2}O_2 \rightarrow CO_2 \qquad\qquad \Delta H = \text{-111 KJ/mol} \qquad R5$$

$$CH4 + 2O_2 \rightarrow CO_2 + 2H_2O \qquad \Delta H = \text{-890 KJ/mol} \qquad R6$$

Text Box 4

In R3, the ΔH here is positive, which means that every molecule of H_2O that is inside wood will _absorb_ 41 kJ of energy when it changes phase to steam. Not only this, but to merely raise the temperature of water requires an input of energy, so that before it boils the moisture in biomass will be absorbing energy to get it to the boiling point. Think of the energy input needed before a pan of cold water begins to boil. This pan will continue to just absorb energy without any observable effect until it reaches a temperature when boiling occurs. And it gets worse, for even when the H_2O is in its steam phase, if that steam is trapped in a closed container (like a gasifier) it still continues to absorb energy for every °C increase. The energy that is put into a substance and therefore the chemical energy that it contains is called its "sensible enthalpy", which is _additional_ to the energy involved in it changing phase.

Although of an exceptionally high exothermic nature when combusted, hence high calorific value, CH_4 is produced in low quantities (in the range of 0-5%) in a gasifier for reasons related to intrinsic molecular properties (31, 38). It therefore contributes little to the overall gas calorific value.

With certain reactors, pyrolysis can be "adjusted" by altering the heating rates so that the amounts of liquid or gaseous product can be changed. The liquid product is merely the condensed tars, whereas the gaseous product will be smaller molecules such as CO, H_2, CH_4, and traces of other very low weight hydrocarbons. If this product is to be sent directly to a combustor, the liquid hydrocarbons add to the calorific value of the mixture and therefore the process can be efficient. This is usually the fate of updraft producer gas. However, the pipeline between updraft gasifier and combustor must be heated to up to 300°C because below this temperature the tar molecules begin to condense out. This is problematic for two reasons: 1. It is losing some of the calorific value of the gas as it will never reach the combuster; and 2. The tar sticks to surfaces, and although in relatively low quantities, over time it leads to major clogging and corrosion.

5.1.4. reduction zone

It is very important to observe on Figure 9 that the reduction zone is converting a molecule of carbon dioxide to one of carbon monoxide by reaction (R7). A chemical reaction of this type, where oxygen is removed, is called a "reduction" reaction, hence the name of this zone. Since the optimisation of gasifiers is achieved by having a large concentration of unoxygenated molecules (so with greatest combustion potential, *i.e.* "calorific value") reducing reactions are obviously to be encouraged. Relatively high temperatures are needed in the reduction zone as the desired reactions here are highly endothermic. This is achieved by being adjacent to the combustion zone and so acquiring heat (through radiation) but not oxygen.

By looking at Figure 9 it can be seen that the updraft gasifier is only able to "reduce" CO_2 because only carbon is an available reactant. This limitation occurs because the reduction zone is positioned upstream of the pyrolysis zone. If it were somehow located *after* the pyrolysis zone, then it could theoretically "reduce" other molecules such as hydrogen (R8), and even the gaseous tar molecules thereby both purifying and cleaning up the gas.

$C + CO_2 \rightarrow 2CO$ (Boudouard reaction) ΔH = +172 KJ/mol   R7

$C + H_2O \rightarrow CO + H_2$ (water gas reaction) ΔH = +131 KJ/mol R8

Such a gasifier exists, and it is with this that the remainder of this book will focus. Its name has previously been mentioned - it is called the "downdraft" gasifier - and the detail of optimising the reduction zone is central to successful operation of it. Previously only simple steel cylinders were considered, but the downdraft gasifier has internal adaptations.

A schematic of the downdraft gasifier is provided in Figure 10. As can be seen, it has the same four zones created by temperature in which the same chemical reactions occur; but in a different arrangement. The drying and pyrolysis zones are in the same position as with the updraft gasifier, but the difference is that the air enters in the mid-section. Because this creates exothermic combustion reactions, the mid-section is the location of the combustion zone. Here, below the air entry points, the reactor narrows into a "throat" constriction (cf. also Figure 8), and this is a very important adaptation. Moreover, the reduction zone is now downstream of the combustion zone, and all the gases pass through this zone en-route out of the gasifier.

Consequently a downdraft gasifier creates a gas with higher calorific value and at the same time removes the tars which would otherwise deposit in or prior to the engine. As R7 and R8 consume the carbon, the char diminishes in size in a downdraft gasifier until it is so small that it falls through the supporting grate rather than completely burning out as in an updraft gasifier (Figure 11).

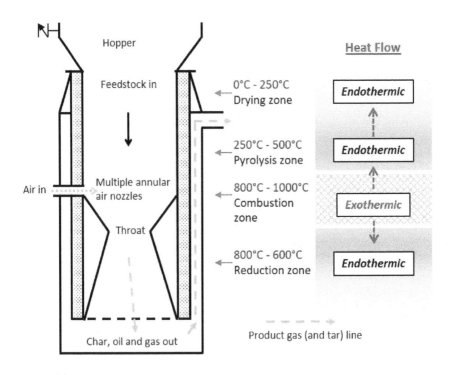

Fig. 10: downdraft gasifier schematic. T/C = recommended location of thermocouples. For details of the nozzle configuration and throat dimensions see The Downdraft Gasifier and Appendix A.

Observant readers may notice by comparing Figure 9 with Figure 10 why the downdraft gasifier is less thermally efficient than the updraft. The drying zone is upstream of all the other zones, which means that the released water vapour must pass through them all. Some of this moisture will indeed be beneficially reduced in the reduction zone, but as it passes through the combustion zone it will impose a parasitic heat demand, which therefore makes initial feedstock moisture content an important consideration.

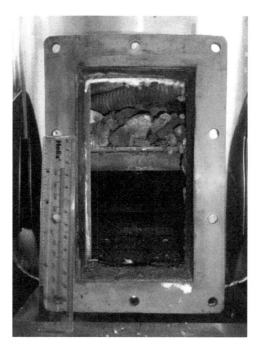

Fig. 11: view inside the gasifier reactor showing the charcoal at the base of the reduction zone supported on the ash grate. The ash grate has recently been cleaned out in this photograph.

Finally, the "water gas shift" reaction:

$$CO + H_2O \leftrightarrow CO_2 + H_2 \text{ (water gas shift reaction)}$$
$$\Delta H = -41 \text{ KJ/mol} \qquad R9$$

Be aware of the similar, but not to be confused, terminology between the water *gas* reaction (R8), and water *gas shift* reaction; and the fact that R9 is mildly exothermic (so that it gives off heat when it moves to the right) as opposed to the other reduction zone reactions which are endothermic. The water gas shift reaction is reversible, and for any reversible reaction, by Le Châtelier's principle, at equilibrium, endothermic reactions are pushed to the right by higher temperatures, but exothermic reactions are pushed to the left (see Text Box 4). The water gas shift reaction therefore begins to favour its reverse direction at temperatures above say 200°C, producing favourable CO, but also water vapour. How this reveals itself during gasifier operation is discussed in the next section.

5.1.5. conclusion

The information in this chapter has been predominantly theoretical. But it was necessary to explain this so that the reader can understand the practical aspects of gasifier operation that will be described in subsequent sections. One can operate a gasifier without this knowledge, but temperature, gas composition, types and quantity of tar, soot, ash and char, are all useful indicators for the operator when access to the hostile environment of the reactor is impossible. They, and the information provided in this chapter can be used to optimise the gasifier system and diagnose problems. Reactor chemistry, and optimising the reactor core, is the foundation on which successful gasifier operation is based.

To recap, the combustion zone provides heat energy to drive the reactions (drying, pyrolysis, and reduction) in the other zones. Reduce the air intake in relation to the reactor diameter and there will be insufficient heat supply to create satisfactory pyrolysis or reduction and the result will be a dirty low quality gas. Increase the air intake with respect to a fixed reactor size, and the combustion zone will spread into the gas producing regions and the result will again be low quality gas. Now, because of this delicate balance of sizing, downdraft gasifiers are highly feedstock specific: to a lesser extent this is chemical, due to reactivity and activation energy, but most influential is the amount of moisture that it contains.

Fortunately most biomass available in a region will usually have similar reactivity, and there are ways around both the moisture content and improving feedstock versatility (see Feedstock), but this is an area of great scientific interest, because in theory if gasifiers could be made to accept a wider range of material this would then enable the input of any carbon containing solid waste.

6. Feedstock

"One hundred years of gasification research and commercial applications have clearly shown that the key to successful gasification is a gasifier specially designed for a particular fuel. It is of paramount importance that the physical and chemical characteristics of the fuel do not change significantly." (40)

In theory, anything solid and containing carbon can be made to release combustible pyrolysis gas and leave behind its charcoal framework. Therefore "if it burns, the engine will turn". This theoretical assertion extends to plastics, leather, cotton, etc, hence why people are looking at gasifiers as being the salvation of waste-to-energy technology in place of large-scale incineration and its negative associations. Because gasifiers are also proven at the small-scale, they become even more attractive as one can envisage domestic waste conversion systems much like a kitchen appliance where all organic derived material can be turned into renewable energy.

At present however, this is not a reality. Even with woody biomass, many gasifier failures are caused by a lack of adherence to specified feedstock tolerance values, resulting in poor gas quality. Chemical composition is not the problem, but rather a feedstock's physical properties. As such the challenge seems relatively easy to overcome, and much research is going into this.

This chapter describes the reasons for this and explores the potential for, and obstacles to, succeeding with gasification of other feedstocks, some of which may seem, at face value, to be feasible and highly attractive. Knowing what is acceptable will quickly become apparent to the gasifier operator. But, laziness or lack of care will lead to gasifier problems. It is perhaps because of the seemingly simple nature of the problem that rules which must be followed with respect to what material a gasifier can accept are invariably flouted, either through operator complacency or overconfidence. Diligence, forward planning, and effort with respect to feedstock cutting, drying, and its transportation will ensure that this one of the "three primary aspects of gasifier success" is correct.

In order of importance, the properties of feedstock with respect to the successful operation of a small-scale downdraft gasifier are:

1. Physical shape, size and strength.
2. Moisture content.
3. Chemical composition
     a. Volatile to fixed carbon ratio
     b. Ash content, reactivity, calorific value

Figure 12 shows tree prunings being prepared for use in a small downdraft gasifier. It is a good guide that if secateurs will cut through it, then it is a reasonable gauge of the correct size. Discard the cuttings and use the rest. The useable material will then need cutting into chunks, each about 2 to 4 cm long. For a small throated downdraft gasifier, this is the ideal feedstock. Bark is fine to be left on, and some amount of rot is acceptable too, but don't use overly dirty wood as the sand and clay can encourage melted ash deposits inside the reactor. Chipped wood can also be absolutely fine, but care must be taken to ensure there are limits on the amount of pieces that are out of the correct size range.

6.1. physical properties

Dense packing inside the reactor will result in inadequate transfer of pyrolysis gases, air and heat, with the result being a producer gas overly rich in tar. It is essential therefore that void spaces are present in the reactor bed. To maintain these conditions, a certain size and angularity of feedstock is needed, but also a feedstock that will not disintegrate under heat and attrition. Chopped wood, and material like nut shells have perfect physical properties, because once they have released their pyrolysis gases, and because of the fixed carbon that they contain, they remain the same rigid angular volumetric shape.

Fig. 12: tree and shrub prunings being prepared by separation as a small gasifier feedstock. The small diameter roundwood branches in the foreground will be used as gasifier feedstock.

Reactor beds can soon become blocked if the feedstock has initially a large quantity of "fines" (small pieces among the feedstock such as dust and other biomass fragments) or if fines are generated inside the reactor by feedstock disintegration. This cannot be stopped completely, and a reactor will tolerate some percentage of fines (perhaps as much as 10% if the rest of the feedstock is good), but the main consideration of feedstock management will be to minimise fines as much as possible and also to remove overly large pieces.

There are European Union classifications of wood chip that give varying quantity of fines permissible within a specified size range. A good arborist or wood supplier should be able to provide high quality chip to this standard, although these classifications were not created with gasifiers in mind. Drum chippers (where the wood is pushed into a rotating drum that has metal teeth for shredding) are the cheapest and most common type, used by almost all arborists whose objective is presently to de-bulk their waste material rather than make a feedstock of a standard specific for gasifiers. Cone chippers use a bladed helical screw into which the wood sections are forced with the distance between the blades determining the size of the chips produced. These can create a better quality chip as far as downdraft gasifiers are concerned as they bring consistency to the sizes and generate far fewer fines. However, the cost of these machines is high and they are rare. That said, I have used drum chipped wood that has been of cone-chipped standard, which depends on the setting and the quality of wood. Even the quality of cone chipped wood can be too poor, as the blades need to be kept razor sharp.

If only drum chipped, or poor quality feedstock is available then the screening out of large pieces (often long and slender pieces) and fines will be required. Though simple and cheap, without some labour-saving ingenuity, this task is time consuming. Making a 1 metre square sheet of mesh, standing it at 45° and shovelling the chip onto it for example, is one option for size separation with the useful chip falling under gravity and the fines falling through (see Figure 13 and Figure 14).

A gasifier operator will soon realise what feedstock physical characteristics work and what don't. It is not a disaster if too many fines are created. The worst this will mean is that the reactor will need to be emptied out and the process started up again, with obviously the system shut down and no energy produced. The fines can also be made use of, either as garden mulch, or as filter media in a dry gas filter (see Dry filters). At the other end of the size range, over-large pieces can also be picked out by screening.

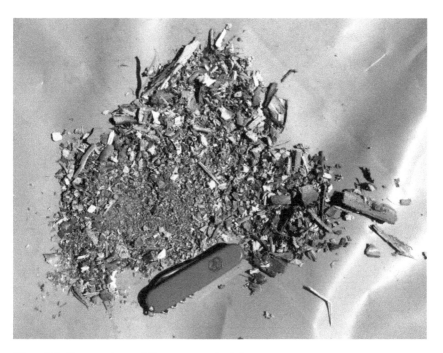

Fig. 13: fines sieved from wood chip

Oversized pieces have less of a surface area for the desired reducing reactions (R8 and R9), which is not advantageous. But the main reason why they should be excluded is because they can cause blockages inside the reactor. Pieces with a rough texture and those which are fibrous are more prone to bind together due to frictional resistance, accentuated by the natural compression that they are subjected to inside large gravity-fed vessels such as the hopper and reactor. When this happens a dome can form and the feedstock is said to be "bridged". The bridge can either remain as a permanent structure resulting in a blockage and then the supply of feedstock being cut off; or the roof of the bridge will collapse leaving a central channel through which the roof enters but through which the sides remain stuck together, ultimately also resulting in the feedstock supply being cut off.

Bridging in the feedstock hopper can be remedied by fitting an agitator into the hopper tank. Bridging inside the reactor can lead to more serious problems as open spaces near the air nozzles cause localised high temperatures and consequently the potential for damage to the nozzle outlets. Mobile gasifiers are more able to rectify these problems due to the vibrations of vehicular motion.

Fig. 14: wood chip from a drum chipper after screening with 1.2 cm sieve and fines removed. Note the large range in sizes. This is poor quality chip for a gasifier. It might be good for the mulch market, and it is the sort of stuff that you can get if you are not careful.

Low bulk density (the mass of feedstock loosely arranged per unit volume) is a contributory factor to bridging and this excludes some otherwise potentially good feedstocks with excellent chemical compositions, such as coconut coir or sugar bagasse. High bulk density is also a desirable property as it means less material need be transferred and handled to create the same quantity of combustible gas molecules. As well as natural form, moisture content also reduces bulk density.

Roundwood billets as shown in Figure 15 have a high bulk density, and because the grain of the wood has only been cut in one place, at right angles to the cross-section, they are robust inside the reactor and the relatively smooth edges give them a very low propensity to bridge. They are the perfect downdraft gasifier feedstock. Pieces such as those shown in the foreground of Figure 12 are available from sustainable land management in most parts of the world.

Fig. 15: roundwood billets - excellent gasifier feedstock

6.2. moisture

The moisture within biomass it is not merely surface wetness. It is present within cells and pore spaces. Freshly cut "green" wood can have up to 50% moisture, and the value depends on the time of year when it is cut. If cut in the dormant winter season, wood will have lower moisture in comparison to that which is cut in the growing season when sap is rising through the plant stems.[2]

Log and branch wood is "seasoned" to remove this internal moisture. Seasoning occurs passively if the wood is left outdoors where it is exposed to free circulation of dry air. The time taken depends on a number of variables such as species, time of felling, air temperature, and extent of air circulation through the wood pile, but very generally, it can take about a year. Although it is not essential for the wood pile to be kept covered, this does quicken up the drying process (39).

For those considering buying any gasifier system, before doing anything else, seek out details of what the feedstock specifications are before committing to purchase. Some systems suggest moisture content that is below that which can be achieved by passive seasoning. Without heat exchangers, the World War Two Imbert gasifiers, were tolerant of 15% moisture or less. This is still a good general guide, and it is definitely achievable with passive drying. Less than about 7% average moisture in chipped wood is difficult to achieve passively without excessive effort. The moisture content of wood chip is visibly apparent and with a little experience can be gauged, but for accuracy, hand-held battery operated probes are widely available and very cheap (Figure 16).

Two plots of output data from a downdraft gasifier operating at identical engine load illustrate the effect that moisture has on reactor temperature and consequently gas quality (Figure 17). After a few weeks in storage the wood chip had a mean average of 17% moisture and this was used for the first test. Another part of the same batch was spread thinly outdoors in the sun to reduce the average moisture to 7%, and this was used for the second. The effect on gas composition was that 21% less CO, and 84% more CO_2 concentration occurred with the 17% moisture content wood. An explanation for this comes from the chemical reactions described in

2 H_2O can also form inside the reactor by chemical reaction (see Section Principles of Gasification), but this is excluded from further discussion here.

Principles of Gasification. As can be seen from Figure 17 the higher moisture content created lower mean temperatures in the reduction zone of 65°C (±15) at the top and 49°C (±16) at the bottom in comparison to the dry feedstock. The moisture (hence temperature) had inhibited the endothermic reducing reactions (R7 and R8). Additionally, the increased H_2 and CO_2 can be explained as functions of the water gas shift reaction, where at higher temperatures, this mildly exothermic reaction begins to favour its reverse direction leading to increased CO and H_2O and lower H_2 and CO_2 (40).

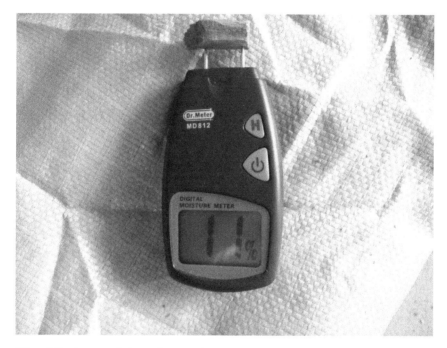

Fig. 16: determining the moisture content of wood chip

Before leaving this sub-section, there are two further things worth mentioning about moisture and gasifiers. Firstly, it may be considered that the heat exchangers in the gasifier shown in Figure 7 will extend the system's tolerance to higher moisture feedstock. But, because this gasifier operates under suction, all the water vapour must still pass through the hot reactor core. As mentioned in Text Box 4, even in its steam phase, H_2O still absorbs energy when it encounters high temperatures, and this sensible enthalpy increases for every °C. Secondly, it is interesting that some steam in the reduction zone is beneficial. In addition to steam being a reactant for hydrogen production through R8, it also promotes the reduction of tar (41). So, if there were some way to extract the steam generated in pyrolysis and allow _just this molecule_ to bypass the combustion zone but then re-introduce it directly to the reduction zone, then that would be a novel and attractive design approach.

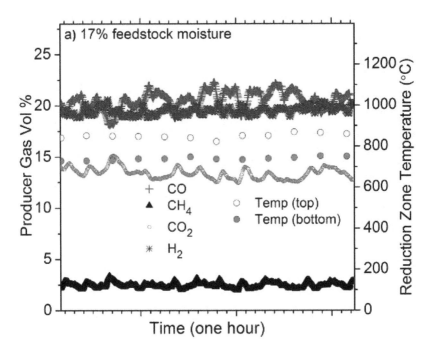

Fig. 17a: effects of feedstock moisture content on producer gas composition and gasifier temperature (17% moisture).

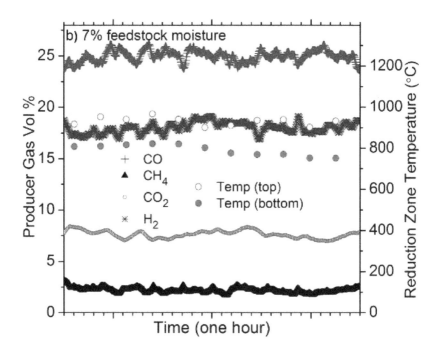

Fig. 17b: effects of feedstock moisture content on producer gas composition and gasifier temperature (7% moisture).

6.3. chemical composition

It is a mistaken belief that the species of tree or shrub used as gasifier feedstock has a major influence on tar production (35). In fact, whether the biomass used is softwood or hardwood will likely have no observable difference on reactor performance.

6.3.1. calorific value

As with producer gas, feedstock energy content can be assessed and compared by its calorific value, which is again determined from the elemental composition of the combustible elements C and H, along with N and internal O_2. Such tests can only be done in a laboratory, but it is theoretically interesting to use the calorific value of the gas produced to estimate the efficiency of a gasifier by comparing this with the energy content of the feedstock. This is expressed as "cold gas efficiency" (Eq. 1):

Cold gas efficiency =

LHV of gas x gas volumetric flow rate

LHV of feedstock x feedstock volumetric flow rate   (Eq. 1)

For all types of wood the calorific value is approximately the same at about 17 to 19 MJ per kg (gross basis or HHV), although moisture content will decrease it significantly. This HHV is often quoted and based on an industrial standard method which includes the potentially recoverable energy from the latent heat of vaporisation of feedstock moisture. Although this can be achieved for example in modern condensing boilers, for gasifiers, there is at present no such means of energy recovery and so HHV should not be used to estimate the energy produced by gas engine combustion. Instead net or lower heating values (LHV) should be used on an ash inclusive basis and with reference to the moisture content of the fuel, which takes the values to about 13 to 15 MJ per kg (26).

6.3.2. fixed to volatile ratio

From Text Box 1 it is apparent that the fraction of volatile gas produced is about three times the quantity of fixed carbon. So, there is clearly insufficient carbon to reduce all the moisture and oxygenated hydrocarbons. This is why void spaces between the char pieces are important for supplementing the gas quality, and why the fixed carbon content of feedstock is important. Most woody biomass will however have similar fixed to volatile ratios.

As an experiment, I once trialled some waste textile that had been shaped and compressed into thick circular disks. In theory, the cotton which it contained would produce a pyrolysis gas comparable to any woody material, and even synthetic fibres would pyrolyse. It was no good though, as: 1, it fell apart inside the reactor; 2, it was so light that it would not feed through, even when blended in a 30:70 ratio with wood chip; and 3, even if it had gone through the gasifier or could have been somehow chemically treated to stop it disintegrating, its fixed carbon content was only 4.6% and another 4% was ash. The volatile hydrocarbons would have produced lots of pyrolysis gas, but there would have been insufficient depth of reduction zone to purify it.

6.3.3. ash

The carbon, hydrogen and oxygen in wood constitute ca. 99.5% by dry mass. The remainder is ash (from mineral elements) and this cannot be pyrolysed or combusted. These minerals remain within the small pieces of char at the bottom of the grate.

The role of minerals in the gasification process is an ambivalent one. They are known to have a catalytic effect on the reactions at low temperatures (42) meaning that the system to engine response time is beneficially shortened when changes in load occur. However, some, in particular potassium (K) and sodium (Na) are fluxing agents: they promote ash melting at relatively lower temperatures. Ash fusion is a troublesome phenomenon for updraft gasifiers as the combustion zone is at the base of the reactor which is also the location of the ash grate. Here, a combination of the high temperatures (ca. 1000°C and above) and a predominantly ash material result in the updraft gasifier having major problems with clinker formation and therefore clogging of the grate (Figure 18). This is particularly so when these gasifiers are used for rice husks that are rich in silica (40).

For a downdraft gasifier operating at normal temperatures (e.g. at or below 950°C), clinker formation is unlikely to be a problem. Firstly, the temperatures are just that bit too low, and in the high temperature regions, the ash is less concentrated as it is still combined within charcoal. Secondly, all woody biomass has similarly low concentrations of ash; this being five to ten times less than contained in leafy shrubs or material like straw and rice husks (26). Small quantities of fused deposit are not uncommon, and these can be cleaned out. It should only be a concern if it becomes excessive, which would indicate to the operator that their reactor core is too hot and/or the feedstock is not right.

Fig. 18: fused deposit extracted from the combustion zone of a gasifier. Size = 2 cm across.

6.3.4. feedstock reactivity

The science of how substances react is called kinetics. Different species of biomass react at different rates and require different conditions of temperature to promote the desired reactions, which need not concern the small gasifier operator as it is enough to know that certain types of material will be consumed in the reactor faster than others. This affects how quickly it is used up and also how quickly the reactor responds to variations in load requirements from the engine.

Fortunately, woody biomass thermal decomposition kinetics are sufficiently similar across species as long as the reactor temperature remains within a certain optimum range. Start using novel wood materials however and at lower temperatures, the speed with which the reactions occur across species can begin to deviate. Figure 19 shows this for reaction (R7). Note how the woody species all have similar reactivity, along with (somewhat surprisingly) sugar bagasse. Note also how operating the reduction zone at higher temperatures tends to standardise the reactivity across all species.

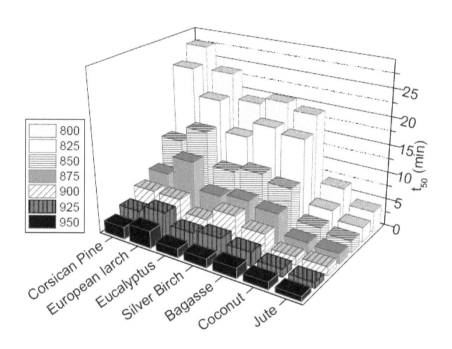

Fig. 19: reactivity of the Bouduard reaction (R7) on different biomass species, shown by time at which 50% of the feedstock is consumed (43).

For completeness, the conditions for which the feedstock is exposed in the pyrolysis zone will affect its reactivity in the reduction zone. This also should not affect the small fixed-bed gasifier operator because the fixed-bed reactor does not permit the variation of pyrolysis residence time or temperature except through the indirect consequence of having a high moisture content or too densely packed bed where insufficient heat throughput occurs.

6.4. additional matters to consider

6.4.1. pellets and briquettes

Wood pellets are made from reconstituted biomass, e.g. sawdust or shredded bio-crop species such as Miscanthus, and they are readily available because they are marketed for biomass boilers. These pellets most commonly come in 6 mm cylinders and can be up to a few centimetres long. This however is slightly too small for throated Imbert gasifiers. Chemically they are ideal as they will pyrolyse just like virgin wood, they will have a good fixed to volatile ratio, they are homogeneous so in theory give a stable operation and gas output, plus they will "flow" easily which means that they would not densely pack or bridge. But, the main reason that they are not suitable is, again, physical. It is certainly possible to make larger and perhaps more angular shapes since pellets and the larger briquettes are made by compaction through a die and press. But, reconstituted biomass pellets, regardless of size, disintegrate too easily under physical stress - they easily snap into smaller lengths and are friable, so the gasifier would need frequent vigorous grate shaking to keep gas throughflow clear. More importantly, they absorb water and when this happens they swell and turn to a constituency of soggy wheat biscuit cereal. This occurs when a closed-topped gasifier is shut down at the end of its operating period (usually at the end of each day). The reactor cools and the moisture released from the drying zone permeates the reactor. For a single run, wood pellets will work for a few hours or more, but try to then start a gasifier up the next day and steamed disintegration will have blocked the reactor bed necessitating a full clean out.

6.5. storage and preparation

If left outside in a bag or a pile, wood chips when green felled will not have sufficient air circulation to allow seasoning to occur, and they may never dry out. The top layer will appear dry, but a few centimetres below the surface the pile will soon be home to fungal growth, decomposer insects, and will be decaying into compost. It is therefore best practice to season the wood first, as with any logwood stacked and interlaced to ensure sufficient ventilation. When cut after seasoning, the biomass can be bagged and stored undercover. If the biomass is cut and chipped "green" (with high moisture content) then it will need spreading out on a tarpaulin or board, preferably under cover (see Figure 20). On a hot day, the top layer will quickly dry out, but it is amazing that even if only a few centimetres deep, the bottom layer will remain damp. Then the chip will need turning, and a whole day could be spent doing this job. The removal of finer particles from the chip at this stage will aid drying. In all cases, some sort of covered storage is required.

6.6. costs

It is not cost effective to buy a small gasifier if you do not have access to freely available wood chip or suitable biomass waste. The reason being that although it is possible to find a local arborist who has waste chipped wood, it is the delivery costs due to the bulk weight that make supply expensive. So, it is worthwhile building up some local contacts. Then it is highly likely that you can acquire wood chip for free, more so if you offer to collect it yourself.

Fig. 20: attempting to dry wet chipped wood

7. the downdraft gasifier

"...gas producers still have the image of a simple stove like energy conversion system easy to design and operate. The present demand is also stimulated by the belief that gasifiers can convert almost any carbonaceous material to useful mechanical and electrical energy. This image of a gasification system is far removed from any reality and in particular the history of gasification has shown that a fixed-bed gasifier providing fuel for an internal combustion engine is a very selective energy conversion system with little flexibility with regard to the fuel it was designed for." (1)

It is now time to look in detail at an actual gasifier. This chapter will explore why it is built in the way that it is and how to ensure that it operates successfully based on knowledge from the previous sections.

7.1. hopper/feedstock container

When the gasifier is started for the very first time, the reactor will need to be filled with lumpwood charcoal. This is proper pieces of charred wood, not the reconstituted material that is sold for barbeques. Once the gasifier begins to operate, the reactor will replenish its own charcoal from the wood feedstock, and will only ever need emptying if maintenance requires access to the reactor core. If there was no method of continuous feeding, the gasifier would not operate for long before the reactor core was emptied. Gasifiers like the Ankur (Figure 6) have a semi-open topped vessel directly above the reactor, or an easy to open lid (Figure 7, Figure 21, and Figure 31). These are both filled at the start of operation and can be topped up when required. The open-topped design works because a small amount of air at this point is neither hazardous nor detrimental to the outputs. During operation, the feedstock falls through the system under gravity, with sloping sides ensuring that the feedstock does not channel. The hopper shown with the Ankur system will last about six to eight hours when fully filled.

Care must be taken if opening the hopper lid for short periods of re-filling. Condensate collects inside, particularly around the lid when the engine is idling or when the system is shut down, for example overnight. This condensate is acidic and so the hopper must be made of corrosion resistant steel. Aluminium has proved to be unsuitable due to ammonium hydroxide content of the wood gases (27). Some hoppers have design features which capture and remove this condensate by having a drain hole at the top as shown in Figure 21. With no condensate drain and a closed topped hopper, the feedstock will be wet at the next start up, and all moisture must ultimately be drawn through the reactor core.

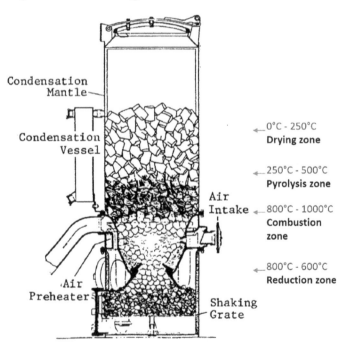

Fig. 21: downdraft gasifier for boats. Air intake and other parts of the hearth are cast in one piece of iron resting upon which is a removable ring to that creates the throat constriction (27).

7.2. by-pass valve, start-up, shutdown and flare

A Gasifier reactor will need lighting to start it going. Some designs have a separate lighting port that can be opened and closed by a screwed cap and which is simply a small diameter steel tube that provides access directly to the combustion zone. Others, such as the Ankur (Figure 6) are lit through one of the air entry points. For stationary systems this is done safely with a propane torch. At the same time there must be active suction to pull air through the system, and this can be either blown in (pressurised, by a fan), or sucked out (aspirated). During normal operation, the latter is done by the engine, but until the reactor has attained the desired temperatures for which the reactions described in Principles of Gasification will occur, the gas will be too dirty and must be diverted elsewhere. For this reason, a bypass valve situated before the tar filtering system (but after the cyclone) can be manually controlled to send the gas to a flare stack. One (or two in series) electrically powered centrifugal blowers are best. They must be ca. 100 to 200 W, and they must be covered from water ingress, but also have vents to allow dirty smoke to escape. These fans can get clogged with tar quite quickly so they need to be easy to dismantle and clean, but like other components in the gasifier system they are robust items. For daily operation, once per week fan cleaning is usual, although this depends on the time left running through the fan. As with other maintenance tasks described elsewhere in this book, cleaning the fans is done to avoid problems before they occur.

The flare stack has a second function which is to burn the dirty gas at startup and shutdown, to mitigate air pollution, and for operator safety, *e.g.* from particulates and CO. With the blowers on, it usually takes about 30 minutes to get a small-scale gasifier up to temperature; longer if it has been stood cold for a long time. By far the best method for identifying when the reactor is ready for conversion to the engine is by installing a thermocouple (temperature sensor), but the flare stack can also assist. A few seconds after ignition, billows of smoke will appear from the stack, followed by light grey smoke and then smoke with a blue-ish tinge reveals that the gas is clean and combustible. Either using a spark or a heating element, the gas is ignited. The flame will initially have the conventional orange colour of combustion gas, and after a minute the ignition port should be capped, but the blowers kept running. The flame should then change in colour from orange to a light transparent, almost luminous blue indicating that the producer gas is good quality and relatively free of soot.

Fig. 22: gasifier flare stack during start-up showing transition to blue flame

7.3. electronics

The electrical blowers, thermocouples, and flare stack ignition systems are all useful additions. They can be run from a car battery which can be re-charged using a conventional alternator and fan belt drive while the engine is functioning. In recent years, the extent to which electronic systems have been incorporated into gasifiers has been one area where attempts have been made at progress.

Monitoring is not difficult, but control is! Thermocouples and pressure sensors are low value, reliable and passive items; and some automation can be successfully applied to both feedstock delivery and ash removal. But total system and internal reactor control is much more challenging.

Certainly this is an area where small gasifier technology can be developed. But at present, electronic control is, depending on your point of view, an advantage or a disadvantage. Greater complexity reduces the simplicity of repair and maintenance which is one of the identified advantages of downdraft gasifiers. It is easy to take a spanner and replace a failed gasket, but if a circuit board breaks the whole gasifier is out of action until the fault is diagnosed. If you cannot get the manufacturer to source you a replacement part, your system will not be repairable unless you do it yourself.

Remember the classic car analogy given in What is a Gasifier. Small gasifiers have not had the same scale of commercial research focus that petrol and diesel engined vehicles have had since the two technologies went in different directions after World War Two.

7.4. engines and generators

One of the benefits that biomass gasifiers have over biomass combustion/steam turbines is that there is no need to use water as a "working fluid". This creates increased efficiency for gasifiers.

There are other aspects that both increase and decrease gasifier efficiency in comparison to other power systems, and these will be mentioned here. Tests have shown that engine lifetime can be as long or longer for those run on producer gas than those run on petrol and diesel, without any additional maintenance (30, 31, 44).

Fig. 23: tar and soot deposits in a producer gas engine throttle valve

A number of current manufacturers can supply producer gas engines. But, pre-existing spark and compression engines can also be converted to accept producer gas. The science of this has not changed and some of the best gasifier publications available discuss how to do it (see 26, 27, 31). Compression engines can either be operated at a reduced compression ratio and with the insertion of a spark ignition system, or in dual fuel mode with the diesel used to ignite the producer gas inside the cylinders. Modern engines with turbochargers and after-coolers however are extremely sensitive to producer gas soot and tar (33, 44).

Engines are set-up to operate on a fuel which has a composition within narrow limits. This set up relates to the calorific value of the combined fuel and air mixture that enters the engine cylinders with each stroke, the *amount* of mixture that enters the engine cylinders with each stroke, the efficiency limits with which the engine can convert this mixture into output power, and the number of combustion strokes per unit time. Although CO has a lower explosion rate than petrol, H_2 has a faster one, and with biomass feedstocks they compensate for each other once the reactor is operating correctly. The bulk of hydrogen in the producer gas comes from R8 (40), and because of this charcoal gasifiers predominantly have CO and very low quantities of H_2 (45). The main consideration is however ensuring that the gas is not sent to the engine before the reactor is hot enough and ensuring that the reactor operates at a steady state.

Since the molecules in producer gas are different to petrol, they will require different air:fuel mixtures. The calorific value of producer/synthesis gas is ubiquitously ca. $5 \leq MJ.m^3 \leq 10$ (31, 38), combined with the differing values of stoichiometry, this gives heating values of producer gas:air at $2.5 \ kJ.m^3$ compared to petrol at $3.8 \ kJ.m^3$ thus resulting in ca. 35% lower power output for a producer gas engine (26). Since the producer gas is a delicate function of reactor operation however, these calorific values will change if the reactor is operating outside of range, and therefore the power loss will undoubtedly be lower than the maximum due to the instability of the process and non-uniformity of gas composition. One way to manage this is by installing an oxygen sensor in the exhaust.

The actual energy created by combustion of producer gas is based on the volume of mixture which can be combusted in an engine cylinder, The volumetric capacity available is fixed by the pressure differences across the system which influence the quantity of gas:air that enters the cylinder prior to the start of the compression stroke. Gasifier systems are likely to have lower pressure supply due to necessary upstream components within the filter/cleaning system and this results in a reduction of about 35 to 20% volume below the theoretical maximum compared to normal engines of 30 to 10% (26). This can be overcome somewhat by a wider air inlet manifold to reduce gas flow resistance. The gas will contain the diluents of N_2 and Ar, but minor CO_2 will also act as an inert gas in the combustion chamber since it is fully oxygenated. The higher the temperature of gas going into the engine also reduces volumetric efficiency, and can heighten the propensity for pre-ignition, hence the dual purpose of the post-reactor gas cleaning and cooling system.

Beneficially, producer gas and air mixtures have octane numbers slightly higher than petrol:air mixtures and consequently compression ratios can be somewhat higher at up to 11:1 before uncontrolled combustion occurs. Smooth operation at ratios up to 17:1 has even been shown to be feasible (30). As a consequence, gasifier-fed engines have the potential for higher thermal efficiencies.

Producer gas and air mixtures combust at a slightly slower rate than petrol:air mixtures, and because of this, for longer operation and higher efficiencies, engine timing (ignition in relation to the piston stroke) should be altered if setting up a petrol or diesel engine for producer gas operation. Experience has shown that timing advancements by 10° to 15° for spark ignition, and also diesel injection advancement of 10° for a producer gas compression engine operating in dual fuel mode (26, 30). See Figure 24.

Variable control can be achieved by having an RPM reader on the engine flywheel which links to a governor that manages the air:fuel intake throttle. Vehicular gasifier systems can be controlled in this way by manual throttling. This range over which a gasifier can operate is called the "turndown ratio" (maximum practical gas generation rate/lowest practical gas generation rate), and relates to reactor sizing relative to the engine.

To meet requirements for variations in power transmission, the gasifier reactor will need to be sized so that it has a rapid response to changing engine load. This is particularly so for vehicular gasifier systems. Reactor over-sizing will lead to better operation at high load, but dirty gas at closer to idle speed. It is not just the size of the reactor, but also internal dimensions of the throat, the distance between throat and air nozzles, and the amount of air allowed in, as will be explained in the next section. There is not much that the operator can do about this if he/she uses the tried and tested

Imbert design, unless they decide to experiment and make major alterations, which is not advised.

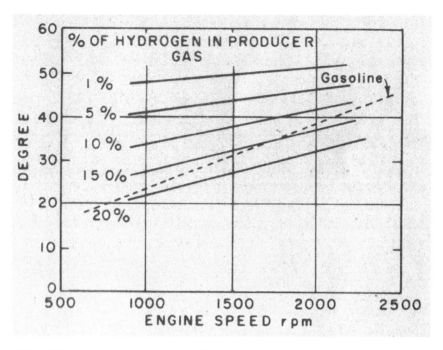

Fig. 24: *recommended ignition advancement (degrees from top dead centre) with respect to hydrogen content of producer gas (40)*

7.5. reactor materials and sizings

It can be seen from Figure 10 in Reduction zone, that the small-scale gasifier reactor casing is twin-walled, with insulation so that heat generated by the combustion zone transfers internally rather than being lost through the external casing. Some larger systems have rigid ceramic insulation on the inside of an iron outer shell, although ceramic is prone to crack because of rapid heating and high temperatures.

As previously mentioned, a key factor in gasifier success is ensuring that free movement of gases is not impaired, for this aids both the preferred internal reaction chemistry and also improves response time to engine load variations.

Consequently, all gas transfer lines should have curved bends to negate pressure drop across the total system. That is, all except the place at the base of the reactor (illustrated in Figure 10). The 90° bends here are to encourage the passive removal of soot particulates entrained in the gas flow. After this, because of tar and condensation, pipes should always be inclined away from the reactor and have a condensate collection point. Easy access and the ability to empty these traps is important.

A combination of pyrolysis gas and moisture make for an acidic environment inside the gasifier. Steel can be vulnerable to corrosion from hydrochloric acid (HCl) (46), but fortunately chlorine in wood is relatively low. Other acidic compounds form in tar however (particularly acetic acid) such that the internal environment can have a pH as low as 2.6 and 3.7 (47). Acid resistant steel (316 or 317) is therefore preferred. The hot regions of the reactor need to withstand 800°C to 1200°C and so the acidic conditions, combined with the thermal stresses which the nozzles and throat are subjected to results in these components being more prone to failure (see Above the Gasifier Throat). Because of this, and partly also for ease of maintenance and repair, it is better not to weld everything, but rather have bolted/threaded connections. The throat can be made to be replaceable by making it from a steel ring that rests in place (see Figure 25).

In less hostile regions, away from the reactor throat, high temperature tolerant gaskets are the best option. Only the reactor core section will experience temperatures above 250°C so some high temperature silicone sealant is very effective between joints elsewhere. One of the main things is however making the reactor easy to access for cleaning or repair. Sadly, many gasifier systems do not provide this, and to fix anything inside the reactor can be an awkward, time consuming, and generally messy task.

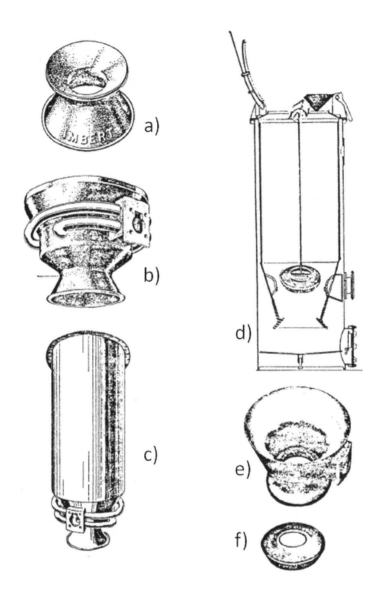

Fig. 25: Imbert style gasifier and spare parts: a) hearth cone of alloyed cast steel; b) hearth cone with air intake; c) complete inner mantle with hearth and air intake; d) how to change the hearth ring; e) complete hearth ring with primary air intake and hearth ring placed inside; f) hearth ring (cast iron). Adapted from (27).

7.6. above the gasifier throat

The old Imbert gasifier design shown in Figure 25, and the more modern Ankur gasifier (Figure 6), have no physical separation between pyrolysis/drying zones, and the feedstock hopper. The GEK system (Figure 7) is different. Here the hopper is offset from the top of the pyrolysis zone by an auger which transfers feedstock from one to the other.

There is little else to describe about the drying and pyrolysis zones. These two zones experience a decreasing gradient of temperature the further their distance from the air nozzle entry points. There is little need for added complexity to internal configuration here and the drying and pyrolysis zones are merely contained within a straight-sided cylindrical chamber, although one that is optimised by being tall and slender. This elongate shape increases residence time, therefore ensuring that drying and pyrolysis is fully completed, and also encourages the feedstock to flow satisfactorily. Further down this cylinder, the combustion zone configuration is unique and it is the correct maintenance of this that is crucial for production of tar-free gas.

As the feedstock progresses along its journey towards the combustion zone air nozzles it will eventually have lost about 70 – 80% of its mass, but importantly it will be of roughly the same size and shape. The evolved gases (pyrolysis and moisture) are being pulled through by the bottom aspiration and so all will pass through the high temperature region of the air nozzles and through the narrow throat. The solids flow is less dynamic but still moves gradually downwards each time the bottom grate is agitated and small char pieces fall through the grate, loosening the packed column above. This means that at the combustion zone, both the char and the pyrolysis gases are viable combustion reactants, and some portion of them must be combusted thereby sacrificing a quantity of the product gas calorific value to self-sustain the system. Ideally, to reduce the amount of tar in the outlet gas, and to preserve the size of the char pieces for the reduction zone, it is preferable that it is the pyrolysis gases that are burnt, instead of the char.

How this is achieved is now explained, when probably the most important aspects of the small-scale gasifier are discussed: the reactor core air nozzles and throat.

7.6.1. importance of air nozzles and air entry

It can be seen in Figure 10 that the air coming in is routed down by the high temperature zones before rising back up again and entering the reactor. This is for non-contact heat exchange, to increase gasifier thermal efficiency by pre-heating the air. It is shown more clearly on Figure 25 (b, and c) how the air is made to circulate around the throat section after entering the manifold. A section of flexible steel hose is also visible among the char pieces in the reduction zone in Figure 11.

A view looking down into this reactor is shown with Figure 26 with the five equidistantly spaced air inlets marked by arrows, and also the single external inlet. This external aperture has a free-swinging butterfly valve through which the air enters a manifold.

It is not visible in Figure 26, but Figure 10 and Figure 27 illustrate how the throat is lower than the annular nozzle array. The ratio of these dimensions is absolutely crucial and should not be changed (See Appendix A).

"Tuyeres" is an old French term for gasifier air nozzles and it is still found in common use. These tuyeres are in an annular arrangement with the direction of their air jets all pointing to the centre of the reactor. What this specific design of air entry is for is to ensure that there is an evenly distributed blanket of high temperature through which the pyrolysis gases must pass. The air flow and subsequent heat pattern created by the tuyeres should not compete, but interlink. Then, the correct lobate shape of air entry will evenly distribute a region of high temperature, as shown in Figure 27b.

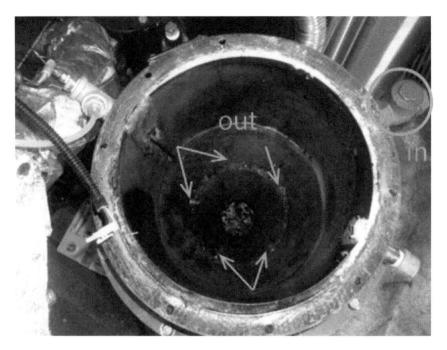

Fig. 26: cross-section looking down into a Power Pallet mk 4 Imbert style gasifier, with air nozzles and air manifold inlet marked.

Tuyere aperture and maintaining its integrity is extremely important. The wrong choice of nozzle, or slight damage to a nozzle can result in a cold spot which will let tarry pyrolysis gases pass through. Narrower tuyere apertures produce a smaller, more penetrating jet of air, but they are more prone to blocking, and will not spread the high temperatures as effectively, but too wide an aperture and the air will not penetrate thus leaving a cold spot in the reactor centre, through which cold and dirty gas will pass. Length of the nozzle can be used to vary the diameter of the nozzle ring. For all these reasons, nozzles which can be replaced are preferable for reactor tuning or for damages. It also does not follow that simply increasing the air inlet velocity will increase oxygen penetration and for this reason, success with gasifier scale up is limited (40).

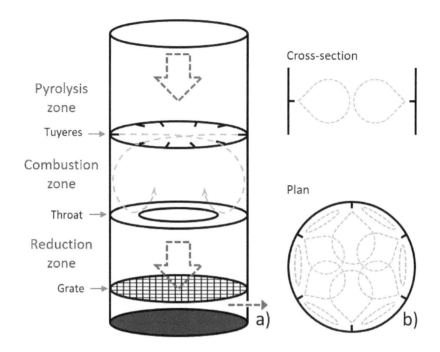

Fig. 27: air patterns and temperature fields inside the combustion zone of a downdraft gasifier. Red arrows in a) show feedstock and pyrolysis gas flows, green dashed lines show air patterns.

7.6.2. importance of the throat

The throat restricts the area through which the gases have to pass, and the design aim of this region between nozzles and throat is to create and maintain a uniform field of high temperature. The design sizings of distance between nozzles and throat (Appendix A) should be strictly adhered to as experiments have shown that if they are varied, the gas quality decreases (26). Undoubtedly this is caused by cold spots and access through the gasifier of unburned pyrolysis gases. I also have an unproven opinion that vortices in the region between nozzles and throat are established by these specific distances. I have shown this in Figure 27a. As tar-laden gas is passing through this section, the vortices increase gaseous residence time in the hot region, and the longer that the pyrolysis gases are able to swirl around here, the more chance that they will either burn or crack. This theory also explains why excessive fines create tar as they stop this circulation. A vortex conjecture similar to this was earlier proposed with respect to a central vertical air inlet and offset nozzle apertures (48).

The central vertical nozzle array is just one of a number of other reactor core configurations that have been tested, particularly on larger diameter gasifiers to try and compensate for the width and to cover the mid-section. Others include a series of repeating vertically positioned annular arrays rather than just the one, with the tuyeres offset with respect to the array above. Then there are "dual fire" gasifiers, that have an option to pull in under-grate air (a remnant from coal and charcoal gasification), as well as air from above, *and* also through the centre. Some, which have been used for rice husk gasification, have a continuous slot around the circumference instead of tuyeres. These complexities may enable the greater accommodation of different feedstocks, but none have had notable success in this regard. Setting up the correct air distribution is made more difficult by these multiple air entry points, and it can lead to incorrect temperature balancing between the zones (40). Plus there is one more nozzle to get damaged, and, with the vertical pipe, also internal access is restricted.

7.7. below the gasifier throat

The purpose of this region is to promote the Boudouard (R7) and Water Gas (R8) reactions which occur on the surface of the char and therefore use carbon as a reactant. In this respect it is beneficial for the reduction zone to be deep. However, an overly long depth of char is more prone to block and then the gas quality will be affected. No data is available on the optimum distances between throat and ash grate.

Because the char in the reduction zone must decrease in size by reactions R7 and R8, the small pieces must ultimately be removed otherwise they will cause pressure build up/restrict the flow of gases, and also inhibit radiative and convective heat transfer. Having a grate which can be agitated, rotated, or scraped perhaps by a rotating arm on a vertical shaft (as per the Ankur gasifier in Figure 6), either manually or by a timer, is an essential inclusion in the design. The grate shown in Figure 11 has two identical grilles, the lower one is fixed and the upper one is controlled by monitored pressure sensors which activate when pressure inside the reactor reaches a programmable set point. For other designs see (1, 40).

Finally, assuming that the reactor is operating with adequate suction to pull away the product molecules as they are created, the main limiting factor in the reduction zone is temperature. High temperature must be transferred to the reduction zone by radiation from the combustion zone, and _not by allowing air to enter directly_. Air would create combustion of the char and therefore lessen the volume of the reduction zone. Air leaks at the base of the reactor do cause this and they can often be so small that their effects can only be discerned by a change in the gas quality. It is surprising how much effect a small hole and/or crack can have. For example, I visited an installation with two identical small gasifiers operating simultaneously and in parallel where both were supplied with good quality feedstock, and both were working well, with clean gas, along with the condensate being light brown (so only lightweight captured tar). It was explained to me that one of the reactors was suspected of having a pin-hole air leak at the base of the reduction zone because when operating on high engine load (so high suction and so more air drawn in) there was a very faint red glow coming from this lower section. It was then observed that this gasifier, in comparison to the other which was functioning normally, was producing a tar-rich gas. The explanation was that the very small air entry point was causing two detrimental effects: 1, it made for higher temperatures and so more refractory tars (see Controlling Tar and Soot Formation in the Gasifier), and 2, by reducing the size of the reduction zone, there was less char available to purify the gas.

7.8. operator

Last but not least, it was mentioned at the outset of this book that a competent operator is one of the three things that are needed for a small-scale gasifier to function correctly. The importance of this "component" is therefore more worthy than a short discussion under a sub-heading.

Unless the gasifier is being used as a communal enterprise and/or teaching tool, anyone thinking of buying one must appreciate that they need a little practical aptitude or the will to acquire it. A good set of tools will be needed. A vehicle maintenance socket set will do, along with common hand tools such as pliers, hammer, saw, screwdrivers, etc. In addition to which there will be the need for alcohol solvent such as methylated spirit for cleaning, along with various common materials: bottle brushes, cloths, dust masks, tape, high temperature sealant, replacement nuts and bolts, shovel, perhaps some form of large sieve for biomass screening, oil and coolant for the engine. Obviously if you are building your own gasifier then you will need welding and sheet metal fabrication equipment.

Assuming therefore that this book provides the knowledge, then motivation and diligence to the task are as crucial as anything previously described. This applies to feedstock preparation and regular servicing. A summary list of routine tasks is given in Appendix B.

8. controlling tar and soot formation in the gasifier

"The performance of the system, on occasions, was excellent with little tar content and on other occasions, for no apparent reason, produced reasonable amount of tar. This behaviour was traced to the structure of the bed with varying fuel chip sizes and moisture content." (44)

If you are thinking of buying a gasifier and the salesperson says that it doesn't produce tar, then don't go any further. Despite some claims to the contrary, <u>all gasifiers produce tar</u> and although its formation cannot be completely stopped, it can be managed. Many gasifier owners throw effort and money into the gas cleanup system with things like wet electrostatic precipitators and oil scrubbers, but still fail.

In lay terms, tar is a brown or black sticky, aromatic, highly viscous substance. It also traps soot particles entrained in the gas flow so further increasing the magnitude of build-up. There isn't much tar or soot produced in a gasifier (see Principles of Gasification and Gas Cleaning – Tar, Soot and Ash Control), but over time, just as a dripping tap can create rot and internal damage, this build-up of tar and soot fouls pipework which then requires excessive maintenance to clean it away. Soot is carbon and solid phase (therefore cooled) tar. Soot particulates can be very small (less than 2.5 of a micron). These smaller particulates cause respiratory problems in humans (see Safety and they can pass through even very fine filter systems causing engine problems also.

The production of tar and soot is dependent on the type of gasifier, the feedstock type, feedstock size and composition, reactor temperature, overall residence time and heating rate of feedstock inside the reactor, along with the dynamics of the reactor bed which affect gas and feedstock transfer. All of which explains why getting the reactor core right, ensuring that the feedstock is correct, and having an operator able and willing to oversee all this, overrides considerations about the gas cleaning system. Things get more complicated because different types of gasifier (and variations in operating condition) affect the form that the tar takes and therefore whether it is easier or more difficult to capture. An updraft gasifier closely coupled to a combustor (e.g. boiler) will not have tar issues despite relatively more tar being produced. Tars are a high energy fuel, so it makes sense to combust them with the gas. But if you want the versatility benefits of turning an engine with producer gas then you need to reduce the quantity of tar and soot in the gas. To do this it helps greatly to know how tar and soot is formed for controlling tar production in the reactor. Attempts at gas cleaning will otherwise be futile.

Tar is a generic name to describe a mixture of condensable hydrocarbon molecules, (such as benzene, naphthalene, and other polycyclic aromatic hydrocarbons) which are within the producer gas as it leaves the reactor (see Text Box 5). The tar molecules are complex and various. They change phase between gas, liquid and solid at temperatures between $15 \leq °C \leq 300$. So, as the gas comes out of the reactor at ca. 600°C and enters the engine near to 30°C, somewhere along this gradient of temperature the tarry gas molecules condense out and stick to metal surfaces. The temperature at which these molecules condense is called the "dew point", and as the dew point varies across different gas mixtures, there can be a layer of tar fouling across all regions of pipework.

Even when the tars have condensed, while the reactor is operating and transferring heat through conduction and convection of the hot gases, the tars can stay liquid and this liquid flows to the lowest point. When the total system is shut down, the tar cools and some of it solidifies. (Figure 28 and Figure 29).

Fig. 28: tar visible on gasifier steel components

Fig. 29: solidified tar from a gasifier reactor top access cover.
Diameter of item shown = 15 cm.

Tar does have functional uses, as it will burn, and it can be used to make chemicals as an alternative to fossil fuel synthesis. It is also called "bio-oil", and has been made into products such as asphalt and creosote.

8.1. tar formation

Understanding of tars and how they form, is one area of gasifier engineering that has not stood still over the last thirty years. The several hundred different hydrocarbons that can constitute tar have been classified by the temperature at which they are formed.

From simple pyrolysis between ca. 250°C to ca. 500°C, "primary tar" is created. As its name suggests, this group of tars is the first to evolve, from the thermal decomposition of cellulose, hemicellulose and lignin. Primary tars are relatively low weight hydrocarbon polymeric (C_2-C_8) molecules such as levoglucosan, hydroxacetaldehydes, furfurals, and methoxyphenols (49). They are generally easier to mitigate once they are in the gas stream, because they are smaller and less refractory. Updraft gasifiers, although they produce more tar, release mostly primary tar, if the temperature is kept moderate. Secondary tar is synthesised (thermally converted) from the primary pyrolysis products particularly above 500°C and comprises (generally C_5 - C_{18} polymers) phenolics, olefins, and aromatics (49, 50). In a pyrolysis retort, the process can be stopped at this stage, but gasifiers go to higher temperatures and here, with further heightened temperatures (above ca. 800°C) and longer residence times, another class of tars form by polymerisation of primary and secondary tars (51). These are longer, heavier polymers, called "tertiary" tars and they include polycyclic aromatic hydrocarbons (PAHs) without oxygen substituents (generally C_6 – C_{24}) such as "condensed tertiary": benzene, naphthalene, acenaphthylene, anthracene/phenanthrene, pyrene; and "alkalised tertiary": methylacenaphthylene, methylnaphthalene, toluene, and indene (49, 50). Downdraft gasifiers produce much lower quantity of tar than updraft gasifiers but more secondary and tertiary tars. New tars form as temperature increases and

Text Box 5

Hydrocarbons are molecules that contain hydrogen and carbon. They can be single molecules such as methane (CH_4), called monomers. But some of these single molecules can join together with other organic (those which contain carbon) molecules in long chains, and are then termed polymers. Polymers comprise most of the molecules in living tissue such as carbohydrates and proteins. Organic polymers are repeating units of carbon and hydrogen, from the small, such as ethene (C_2H_4), and ethane (C_2H_6) up to very long chains. Each carbon atom can bond in four places and each hydrogen can bond once (as the methane molecule), but the carbon can bond to another carbon either with a single bond (these are alkane hydrocarbons) or double bond (alkene hydrocarbons). Hydrocarbons can also form in ring structures (such as benzene) and these are termed "cyclic" or "aromatic" because this group of molecules usually have a distinct smell. This is what gives campfire cooked food its flavoursome taste, but aromatics are often also carcinogenic.

Polycyclic aromatic hydrocarbons (PAHs) are longer chained molecules with numerous aromatic structures. PAHs are not just exclusively hydrogen and carbon. They can include "substituent" elements such as chlorine, nitrogen, of bromine in place of a hydrogen atom. Dioxins, are chlorinated hydrocarbons. Cyclic molecules with substituent molecules are called "heterocyclic".

Adding to the difficulties of removing tar, some hydrocarbons are soluble in water and some are soluble in oil. Solubility is a function of atomic configuration, with water soluble molecules being "polar", although the reasons for this are outside the scope of this book. Polar hydrocarbons therefore collect in condensate, and if this gets into watercourses it can be harmful to aquatic organisms.

these new tars are longer chain, heavy, refractory molecules that are very difficult to remove from post-processing stages. So, just aiming for as high a temperature as possible inside the gasifier is not the right approach, and managing tar is not as straightforward as it first seems.

There are chemical analyses that can be used to precisely identify the molecules in gasifier tar. The producer gas can be sampled and there is a standard method for doing this (52), but the recommended procedure is laborious, messy, and not suited to in-field use. Alternatively tar deposits can be collected and sent off to analytical laboratories. Both necessitate analysis by Gas Chromatography-Mass Spectroscopy (GC-MS), which will not be available to the low-impact user. Even then, the GC-MS cannot detect all tars. Based on the types of tar identifiable by GC-MS analysis, and by how the producer gas mixture behaves in downstream components, a different tar classification has been proposed (Table 1).

The class 1 and 5 tars condense out at higher temperatures. The class 2 tars are polar (water soluble) so they should all be captured in condensate. Class 3 tars should not be a problem either as they will enter the engine combustion chambers without condensing. It is the Class 4 tars that are the ones which remain, and in this group is naphthalene, a substance that is usually the highest concentration tar molecule, identifiable in the areas where it condenses and then dries as a yellow granular/crystalline substance.

Heavy tars determine the overall dew point such that even though overall tar concentration falls (with greatly reduced concentrations of class 2, and 3 tars), the dew point increases if there are slightly more class 4 and 5 tars (53). Figure 30 explains this with the results of modelling based on experimental tar analyses from different gasifiers. The Class 2 tar plot closely matches the Class 4, and are slightly hidden behind it. As can be seen, the low concentrations of the heavier tars have dewpoints lower than the greatest concentrations of the lower class tars.

	Description	Behaviour
class 1	Undetectable by GC-MS. Known to exist by the difference between all the total mass of tar and the total detectable by GC-MS.	Believed to be the heaviest tars that condense at high temperatures even at very low concentrations. Non-polar.
class 2	Heterocyclic aromatics. *e.g.* Pyridene, Phenols Cresol, Quinolene. (Secondary, converted from primary tars),	Polar, and highly water soluble. They are converted to higher classification tars at $750\,°C - 850\,°C$
class 3	1-ring aromatics. *e.g.* Xylene, Styrene, Toluene.	Hydrocarbons that are too light to be important in condensation and water solubility issues.
class 4	Light polycyclic aromatic hydrocarbons with 2 – 3 rings. *e.g.* Naphthalenes, Indene, Fluorene, Biphenyl, Phenanthrene, Anthracene	These are intermediates formed from growth of class 2 tars. They condense at relatively high concentrations and intermediate temperatures. Non-polar.
Class 5	Heavy polycyclic aromatic hydrocarbons with 4 – 6 rings. Produced at high temperature by growth of Class 2 tars. *e.g.* Fluoranthene, Pyrene, Benzo-anthracene, Chrysene, Benzo-fluoranthene, Benzo-pyrene, Perylene, Indenopyrene, Dibenzo-anthracene, Benzo-perylene	Condense at relatively high temperature, at low concentrations. Non-polar.

Table 1: tar classification based on (35, 53)

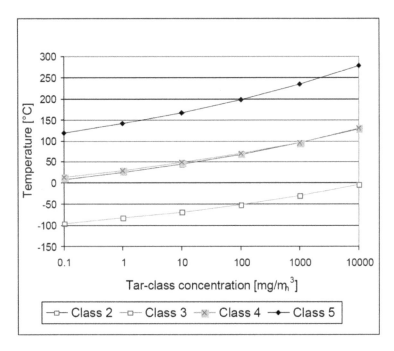

Fig. 30: tar dew points at atmospheric pressure plotted against tar concentration for different tar classes (53).

To conclude, the type of tar that a gasifier produces is therefore indicative of the temperature and conditions that pertain inside the reactor. Due to the impossibility of seeing the goings-on in the high temperature and dark regions, tar identification can be useful information for correcting any deficiencies in performance. By visible appearance, if the tars in producer gas are brown, then these are primary or low weight secondary tars, and the reactor temperature is moderate. But if the tar is thick and black, then these are secondary or tertiary tars, indicating high reactor temperature.

8.2. soot formation

Pieces of soot (or dust) in the gas stream can be more of a problem for gas to engine systems than tar. These are blown through from the gasifier if the base section is not configured correctly or if the waste char and ash is not removed frequently. Soot in the producer gas is not only inefficient (as it is carbon that has not formed a combustible molecule), but it also necessitates measures to clean it up, preferably dry filtration. Size can vary between mm to sub-micron scale. As these particulates are entrained in the gas flow, they can be removed by mechanical separation techniques (see Section Gas Cleaning – Tar, Soot and Ash Control).

Soot is essentially fine particulate carbon. It is the visible smoke that we see from a fire, particularly when first lit or with damp wood, and it is the blackening that we see above the fire and in chimneys. When soot-laden gas is ignited, the soot burns incandescently and this is the cause of the yellow colour of a flame, along with its source of radiant heat transfer. So, oxygenation reduces soot formation because it causes soot particle burnout; indeed oxygen is often added to industrial plants as a method for thermal tar cracking.

Soot production is known to be connected with the mechanisms of tar formation and destruction, being formed in high temperature environments, from long-chain tertiary tars (49, 50). Some studies have suggested that smaller feedstock pieces produce more soot than larger pieces and also that softwoods produce more particulates than hardwoods (26). This may however be due to the effect that packing has on gas throughflow in the reactor bed and hence tar and soot formation dynamics rather than any intrinsic property of the feedstock.

8.3. the effect of biomass composition

Indirectly, feedstock moisture content will affect tar and soot production in a gasifier because of how it lowers core reactor temperature (see Combustion Zone). The amount of power being demanded by the engine (*i.e.* gas production rate of the reactor) will also affect the internal reactor temperature. A gasifier may tolerate relatively high moisture feedstock if it is operating at or close to its maximum rated capacity as then the engine will be working faster and so the air intake will be greater (hence reactor core temperature higher). Similarly, for lower electrical outputs, the reactor temperatures will be lower.

The varying concentrations of lignin, cellulose, and hemicellulose, have little effect on the practicalities of removing tar from producer gas downstream of the reactor. "Lignin produces slightly more tar than cellulose, but the tar dew points are the same" (35), meaning that regardless of the component, the difficulty in removing the tar is not affected.

Fixed to volatile carbon ratio can affect tar production by the extent of reduction zone created. Reactivity of the char also affects the speed with which the reactions occur and therefore how fast it is being consumed.

8.4. the effect of biomass size and shape

With respect to tar production this relates to maintaining a loosely packed reactor bed and reduction zone so that the throughflow of heat and gases is optimised. It allows for spaces between the char pieces where the longer chain hydrocarbons in the pyrolysis gas will experience radiative heat and be broken down.

8.5. the effect of temperature

Temperature inside the reactor is the major influence on tar production. The following extract is based on extensive tests with numerous pyrolysis and gasifier reactors over a 19 year period (35):

"The temperature has the most marked effect on tar amount and composition. A higher temperature promotes polymerisation, resulting in compounds with a larger number of rings. The total amount of tar decreases, but the concentrations of class 4 and 5 compounds increase. Because the heavier tar compounds have lower vapour pressures, the tar dew point rises with the gasifier operating temperature. Increasing the gas residence time in a hot

zone has a similar effect too but smaller than increasing the temperature".

Very high temperatures will crack all tars, but as the above paragraph indicates it is mistaken to go for higher temperatures in the hope that this will reduce gasifier tar problems. The temperatures in a gasifier will never be sufficient for this:

"The rate of thermal cracking is such that high temperatures are required – in the order of 1200°C or higher (also depending on the residence time at high temperature) – in order to break down enough tars so that the remaining fuel gas can be used problem-free in a downstream device such as a gas engine, gas turbine or catalytic synthesis processes." (50).

"Converting tar completely to gas requires greater than 1,100°C without catalyst." (54).

"Temperatures lower than 1,000°C - 1,100°C are inadequate for thermal tar cracking and elimination." (55)

In summary, without catalysts, tars do crack by just high temperature alone. However, it is not straightforward. The oxygen containing class 2 tars convert between 700 to 850°C, but the heavier, non-oxygenated compounds need much higher temperatures of 850 to 1200°C (50). The heavier tars also *form* at high temperature, so by just aiming for this, what will happen is that the tars will likely be in lower quantity, but more difficult to remove. High temperature re-polymerises the simpler tar molecules that come from pyrolysis to create "heavy tars" which condense at high temperatures and are black, viscous, and refractory. In addition to this, the unoxygenated conditions create soot. Thick black tars and lots of soot indicate high temperatures and poor reduction.

8.6. residence time and heating rate

For a downdraft gasifier, the heating rate is likely to be uniformly slow. Heating rate may increase slightly with changes in engine speed which in turn are related to generator load or power demand. It is interesting to note that fast pyrolysis is a standard method for tar *production* because fast heating of biomass in unoxygenated conditions (at short residence time) favour liquid products. This process is often called "flash pyrolysis". So, there is a gradual spectrum with regard to pyrolysis heating rate, with the slowest heating rate favouring gas yield, and highest heating rate favouring pyrolysis oil (tar). This is mentioned here in the main for completeness rather than because it is a variable that may be altered inside a small-scale downdraft gasifier where the heating rate will change little.

Tar will also be produced in higher quantities when subject to shorter residence times. As was described in Above the Gasifier Throat it is beneficial to retain the pyrolysis gases for as long as possible at reasonably high temperature in order to crack tar. Residence time at high temperature can actually lead to tar synthesis, similar to, but with a lesser effect than, single high temperatures exposure (35).

9. gas cleaning: tar, soot and ash control

For producer gas supply to an internal combustion engine, values given for acceptable concentration of tars and particulates are suggested at less than 10-15 mg.Nm³ (31, 40) up to a maximum of 50 mg.Nm³ (1, 55)[43]. For this to be achievable, the raw output from the gasifier should be below 5000 mg.Nm³ (31). However, some engines are better at tolerating tar and soot than others (see Engines and Generators), and some operators merely manage their system by accepting a more frequent regime of cleaning the engine components.

A gas cleaning system serves an additional function of cooling the producer gas. This not only results in the removal of condensate, but also ensures that a larger volume of combustible molecules will enter the engine cylinders (gas volume is a function of temperature), thereby increasing calorific value per unit of fuel drawn in.

Downstream gas cleaning systems will not function if some of the heavier particulates are not adequately removed upstream and so two, three, or more gas cleaning units are common. One other consideration with a gas cleaning system is that suction is powered by the engine intake, and not only does the increased quantity of gas cleaning components reduce the suction strength (pressure drop), but some filters, such as packed beds, immediately begin to cause pressure drop as soon as they are on-line. Sometimes a booster pump is

---

3 N = Normal. Because the molecular volume of gas varies with temperature, this renders any concentration meaningless unless the temperature and pressure conditions are stated. The conventional way is therefore to refer to "normal" conditions of room temperature and atmospheric pressure. 1 Nm³ of gas weighs about 1 kg, so 1 mg.m⁻³ approximates to 1 ppm; and 50 ppm = 0.005%.

deployed in-line, and to ensure that it has adequate power this can be run from a belt drive off the engine.

Post reduction zone gas conditioning starts before the gas has left the reactor. The 90° elbows at the base and the reactor case outlet (Figure 10) take some of the energy out of the entrained flow and help to remove some of the larger (often hottest) particles before they enter the first stage of gas cleaning. This first stage is usually a cyclone, and the presence of extra-large particulates will impair its collection efficiency.

During the Second World War, common gas cleaning systems comprised a cyclone, a wet scrubber, then a gas cooler such as an ambient air fin and tube type, followed by a cold particle trap containing a solid bed of fine fibrous or highly porous material. Figure 31 shows such a system, although the cyclone is absent and would be placed inline immediately after the reactor and prior to the "precipitating tank". Also in this section of the system, and missing from Figure 31, is a T-piece and bypass valve which permits the gases to be re-directed to the flare stack during startup and shutdown. The flare can be positioned after the filter system, e.g. where the blower is positioned, in which case it keeps the fans from being dirtied with tar, but as a consequence passes the dirty gases (at startup and shutdown) through the filter system.

9.1. cyclones

Cyclones are unpowered devices with no need for moving parts. Consequently there are no running costs, and they are durable. They have a vertical cylindrical body with a conical base, at the bottom of which is a collection hopper (Figure 32). They capture particulates by directing the gas along a circular path so that the higher mass particulates (due to inertia) move to the outside by centrifugal forces while the gas continues along its journey towards the engine much cleaner. The gas enters and spirals downward, thereby shedding large particulates. The particulates then fall out into the hopper and, by how the design is configured, the gas

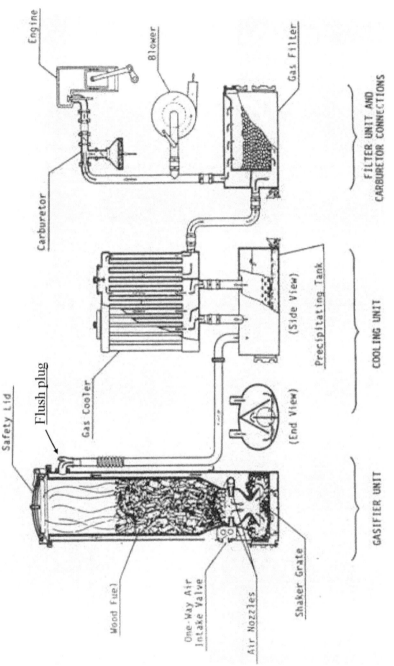

Fig. 31: downdraft gasifier system showing stages of gas cleaning (25).

(minus particulates) spirals back upwards through the centre and onward. This is an effective mechanism of larger particle capture, plus it is compact (usually less than $1/10^{th}$ the size of the reactor). To the author's knowledge it is not possible to buy a small cyclone, but they can be easily fabricated from sheet steel, although the rolling of the steel might need specialist equipment.

Collection efficiency can be improved by adjusting the following parameters:

1. Density, quantity, and diameter of particulates - not favourable to design for this with gasifier operation, and so will be fixed. If the gas is so dirty that large heavy particulates are coming through then the reactor needs to be the focus. To a lesser extent, some form of preliminary gravity separation could be applied, such as sending the gas through a sealed baffled tray.
2. Inlet velocity of gas – this could be adjusted for example by narrowing the cross-sectional area of the gas inlet (*a* multiplied by *b*).
3. Circumferential velocity – this is related to 2 above, but could be increased by having a powered rotor on the inside.
4. Number of gas revolutions – this is number of turns in the helix as gas falls therefore increasing the duration of applied centrifugal forces. It can be roughly equated to the length of the cylinder body.
5. Ratio of the cylinder body diameter (*D*) to the outlet gas spiral (which exits through the outlet diameter - D_e). This affects the risk of particle re-entrainment in the gas flow.
6. Multicyclones. Having a bank of cyclones improves efficiency by splitting the gas to accommodate the same volumetric flow but with smaller inlets and so each having greater inlet velocity.
7. Angle of cone – this causes the gas flow to reverse and the angle keeps the two (down and up) spirals separate. If the angle is too low then the spirals can mix and particulates will re-entrain.

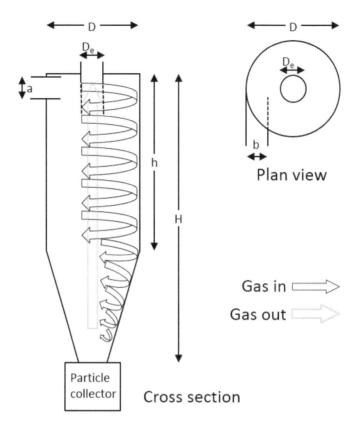

Fig. 32: cyclone particle collector. Usually the first stage of gas cleaning for a gasifier.

The cyclone will not capture smaller particles, and for this secondary polishing stages are needed; however without the cyclone, downstream methods will not function adequately. The cyclone will tolerate high temperature gas flows and so it will remove hot particulates that would otherwise damage fabric filters downstream, and by removing these it can smooth temperatures for downstream applications too. But, cyclones should not be insulated, for their secondary aim is to cool the gas in this section otherwise the next stage in the gas cleaning (usually wet scrubbers) will not condense tars.

9.2. wet scrubbing

Water is effective to a great extent, and also upfront a very cheap and simple gas cleaning medium. It will capture a lot of tar molecules, and some soot and ash too: the polar tar fractions dissolve, while the non-polar tars are captured by water's cooling effect causing them to condense. The non-polar tars, being immiscible in water, along with particulates of soot and ash, form a surface scum which can be scraped off (Figure 33). Tests have reported that water can capture 22 to 39% of non-polar tars (57); so, here we have a readily available substance that can capture all the polar and a reasonable portion of the non-polar gas contaminants: in theory a good choice.

As ever, there is a problem. Water scrubbing just transfers the tar, soot and ash to another medium and then this water will need cleaning up or disposing of. Because of its high phenol content, tarry water can be very expensive to dispose of and it cannot legally just be poured down the drain, although in some countries with less stringent laws, this is done. Because the mixture separates out, settling tanks can be used for stationary installations with the clean fraction of waters then re-introduced to the scrubbing system (58). So, wet scrubbing should be chosen with caution and with a consideration of the need for waste management.

The baffled water tank shown in Figure 33 is a crude method of water scrubbing. The principle is however the same as the "Precipitating Tank" shown in Figure 31 and in greater detail with Figure 34 as the first stage of gas cleaning. Flocculants are sometimes added to this and to other wet scrubbing systems to encourage agglomeration of particles.

Fig. 33: baffled water tank through which the producer gas passes, capturing miscible tars, some immiscible tars (through its cooling effect). a) operational; b) empty. This is a crude tar collecting method.

In the application illustrated in Figure 34, the water scrubbing unit has been optimised by being situated at the base of a vertical cooling tower. Here the water tanks are separated, allowing the second tank to keep relatively clean and therefore permit a higher capture efficiency. The vertical cooling pipes work on the same principle as a car radiator, with fins to increase the surface area for greater cooling. Tar will also build up in the vertical tubes, but having a removable top section allows the inside compartments to be flushed out and easily cleaned.

Vertical wet scrubbing systems also come in a variety of more complex designs and styles (see 59). Some spray water countercurrent to the gas flow in what is known as a scrubbing tower (spray can increase capture efficiency). Another type is a rotating fan (mop fan), through which the gas passes and which contains either a reservoir or a spray of water. These rotating mop fans are compact and could possibly supplement flow velocity, but in my experience they are not effective at tar removal. The fan brushes get soiled in a period of minutes resulting in decreased efficiency and necessitating daily removal by soaking in soapy water or light oil.

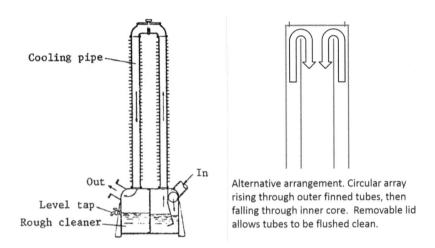

Cooling pipe

Out

In

Level tap

Rough cleaner

Alternative arrangement. Circular array rising through outer finned tubes, then falling through inner core. Removable lid allows tubes to be flushed clean.

Fig. 34: water scrubbing cooling tower. Left image from (27)

Before leaving wet scrubbing, it is interesting to mention the use of oil as a solvent. Zwart, and co-workers, by investigating tar dewpoints, recently devised an oil-based scrubbing system which reduces gasifier tar by 100% for heavy tars and 99% for all other tars, with an overall dewpoint reduced to 10°C (60). This type of system, known as the OLGA, will not be for small-scale applications, as it is designed for large commercial gasifiers. But, it is an interesting concept.

9.3. dry filters

Dry filtration is applied post-cyclone and usually also post-chiller/condenser/scrubber. It is implemented by a series of filtration vessels containing coarse followed by fine material. These are the polishing stages of the gas cleaning train, and various types of media have been used successfully, e.g. textile, foam, small fragments of biomass (fines and feedstock grade material) (26), sand (44), and charcoal (58). Where a series of dry filters are deployed, the coarse filter will usually have a condensate drain. All these filters will capture the fine particulates effectively but, particularly the finer ones, soon become clogged and their efficiency decreases. Consequently they all must be replaceable or cleanable.

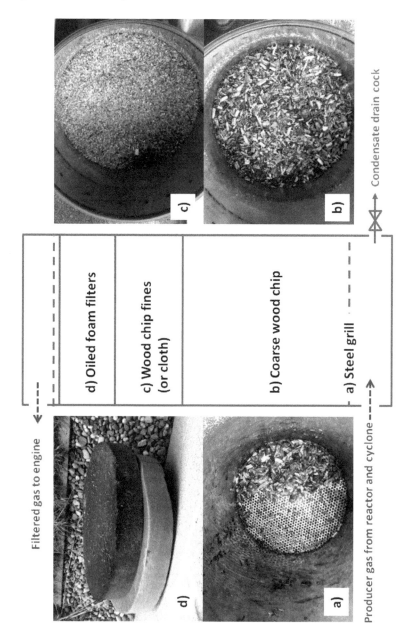

Fig. 35: filter system used in Power Pallet 10 kW gasifier. Note that the dashed line above section d) is also a steel grill to restrain the foam filters.

Dry filters work effectively on a dry gas, but moisture will clog the fabric/medium, so it is best to have removed the moisture at an earlier stage. The alternative is to keep the temperature high and therefore above the water condensation point, but this is risky with fabric filters which can catch fire. Fibreglass, wood or PTFE fabric will however tolerate higher temperatures.

Figure 35 shows the passive filter system in the Power Pallet gasifier by GEK (cf. Figure 7). This is just a steel drum and acts as both filter, gas cooler and condensate collector, importantly without using water. Upstream of this unit is a small cyclone, although there are heat exchangers that also cool the gas, but no wet scrubbing system at all. The lower section of wood chip acts to cool the gas and moisture collects in this region having made its way under gravity to the bottom where it can be removed via a stop cock. The fines used in zone c are leftovers from biomass screening, but old cloth can be used alternatively. Feedstock quality wood chip (in terms of size and shape) material could be used as part of the packed filter and when spent, it could then be used as feedstock in the gasifier. By re-introducing it, the captured tars would be cracked on their second run through, potentially increasing the calorific value of the producer gas and at the same time avoiding costs and labour for tar disposal.

By the time that the gas reaches the oiled filters (zone d in Figure 35), it is relatively clean and dry so that these foam disks are the polishing stages for capturing the Class 3 and 4 tars. Engine oil could be used, but seemingly vegetable oil is likely to be more effective (57). Diesel oil is not recommended as it has components which evolve into the gas and actually increase the tar content (57). Each disk is about 5 to 10 cm thick of open cell-type foam with first coarse and then finer grade. In terms of capacity, this size is about 2-3cm^3 per engine cylinder litre. In the context of vehicles, normal cleaning intervals have been reported at about 1500 – 3000 km (26), and for a stationary system such a filter would need emptying and re-filling roughly once per month. This would obviously depend on the engine load, the filter surface area, and how much operation outside of optimum temperature the gasifier experienced.

9.3.1. cold fabric filters

Baghouse filters are common in industry for capturing fine particulates. They are, as the name suggests, a collection of fabric bags or "socks" grouped together and encased in a box. The bags are gas permeable but impermeable to particulates, and function in the same way as the bag in a vacuum cleaner. In a baghouse filter unit, bags are hung on a rack, and gas flows from below and from outside to in. This traps particulates on the exterior where they build up to form a "cake". The efficiency increases as the cake develops but so also does pressure drop. It reaches a point where the pressure drop is too great and then the cake must be removed. This is achieved easily by the top bar on which all the bags are attached, being activated as a shaker, either pre-set to activate at internals, or connected to a pressure sensor. The cake then falls off and the bag is clean again. After removing the cake, the pressure drop is low but the efficiency is also slightly lower until the cake begins to form and the cycle is continuously repeated

9.3.2. hot ceramic filters

These are used in large scale coal-fired power stations, and work on the same principle as the fabric baghouse filters. They are very effective at taking out particulates, but the gas must be kept at above 200°C and ideally 250°C to ensure that water and tars do not condense out. Small scale versions can be made because ceramic "candles" can be bought relatively cheaply, or high temperature hoses can also be used for the do-it-yourself installation. As with baghouse filters, the flow must be from down to up and outside to in so that the particulates will be captured on the outer surface, leaving the cleaned gas on the inside. The filters will need frequent purging of the captured material which can be done by automated shaking or the blowing of compressed air. Once the purging mechanism is operating effectively they will give completely particle free gas, but they can be easily damaged by this action, particularly where the candles/hoses join. So it is important for tar-free gas that these units are checked frequently for damage, and it is also important to be able to replace one candle/hose if it breaks rather than the whole unit.

The good thing about hot ceramic filters is that they can be easy to upscale, the larger the array, the less frequent the cleaning mechanisms. As a very basic rule, for a well-designed reactor, 1 to $2m^3$ of capture volume per engine cylinder displacement will give a decent operating time before replacement. But, in addition to their vulnerability to being damaged, which will lead to whole system shut-down, the challenge for the small-scale system owner with having a hot ceramic filter is the need to keep the gas hot.

If one of these units is used, then it can be succeeded by a dry cooler/chiller unit to give a most agreeable gas cleaning system. As the particle-free gas is still above the water dew point, it will condense out in a common non-contact heat exchanger, so that the tar will be in the condensate also. This can then be followed by a polishing (cold fabric) or foam filter stage and the whole process will have been achieved without wet scrubbing.

9.4. alternatives

9.4.1. granular packed bed and sorbent filters

Some types of filter incorporate adsorbant material that is able to collect certain components from the gas stream, thereby cleansing it of the substance adsorbed[4]. The gas is passed through a vessel containing this material which can be

4 Adsorbtion ("d") is the capture of a substance on to the sorbent surface. It is different from absorption ("b") which involves taking the captured substance into the sorbent's structure.

either wet or dry. These are, in my opinion not the best solution for a number of reasons, and they will still require secondary filtration.

All packed bed sorbent systems soon become saturated and efficiency drops significantly. They can be regenerated or the whole packed bed can be replaced. Granular packed bed filters contain inert beads, with optimum particle size of about 1 to 2 cm diameter (as smaller beads tend to bind together and clog). Wet systems are flushed with water to cleanse them of solid deposits, often with some detergent which helps to remove soot, but in general these methods give a low overall efficiency for particle collection.

For tar cleaning, lime-based sorbents have been shown as 75-95% effective when used as a secondary reactor for gas cleaning but only in short trials and only when fully calcined (pre-treated at high temperature and under bottled gas flow) (61). Dolomite (magnesium limestone) is one such sorbent that has been used experimentally. Though readily available and relatively cheap, the processing of limestone is environmentally damaging and the potential throughputs of material that would be needed for long-term operation would be costly and unsustainable. Dolomite has inherent friability too, is vulnerable to acid gases, and so is prone to rapid de-activation (61). It also soon becomes saturated and efficiency drops significantly. There will be a short finite number of times when it can be regenerated to bring back some of the efficiency, but soon the whole packed bed of dolomite will need to be replaced.

9.5. electrostatic precipitators

Electrostatic precipitators (ESPs) are sometimes incorporated as a later stage biomass gasifier cleaning unit. They are effective in power stations to capture fly ash, but their cost, size, and electrical consumption does not suit the small-scale gasifier system. Notwithstanding this, my experience of them is not favourable, as in cases where I have seen them used to clean up producer gas from a biomass gasifier, they have not worked.

10. safety

10.1. gas toxicity

Carbon monoxide is toxic by inhalation as it binds to the haemoglobin in blood instead of oxygen. Producer gas contains about 20% CO, but it needs only much smaller quantities, above 0.16% in air, to cause death in two hours; and with concentrations of 1.28% death can occur within 1 to 3 minutes (26). Exposure to lower concentrations is manifest by headache and dizziness. Obviously it is vital therefore not to breathe the gas and to always have a carbon monoxide detector (readily available and cheap) positioned nearby.

During normal operation the risk of carbon monoxide release is unlikely because the system is at negative pressure so rather than leak outward, the producer gas is sucked into the engine where it will be safely combusted. At startup and shutdown, the system is made safe by igniting the gas through the flare stack. The greatest danger from accidental carbon monoxide poisoning comes after shut down, and in particular when repairing or maintaining the gas lines and components during these periods as the gases remain trapped inside not just the reactor and hopper but also the ash grate and gas filters. In this respect be especially careful when changing the filter media, filling the hopper or emptying the ash grate. Not ventilating the unit after a day's use benefits quick start-up on the following occasion by containing the combustible gases inside, but here is where the hazard is created. Either by complacency, ignorance or forgetfulness, someone may forget when checking the feedstock level the next morning resulting in exposure to potentially fatal concentrations of carbon monoxide. This is documented as happening in Sweden during the 1930s and 1940s with most gasifier fatalities associated with people repairing their gasifier vehicles inside a garage during cold weather (26, 27). Because of these risks, gasifiers are best sited outdoors and under cover or in an area that is well ventilated.

10.2. noise

Fixed-bed gasifiers make no noise when operating. The engine will create noise, and for this reason, along with the bulky store of feedstock necessary, small-scale gasification systems are at present too large to be considered as an interior domestic appliance. Even a small gas engine will have noise levels at 1 – 2 metre range of slightly above 80 dB, the daily E.U. workplace exposure limit (my system tested at 85 dB). In a car, the noise is soundproofed to a large extent, by the bonnet, and similar soundproofing is possible for a stationary biomass gasifier engine.

10.3. high temperature, fire and explosion

Operating temperatures inside the reactor are between 800°C to 1000°C for optimum clean gas, but the danger of fire and of burns to the skin is reduced by the reactor being well insulated. The flare stack will not likely be insulated and this will get very hot for short periods, but it should be out of the reach of skin contact. The external casing of the cyclone may get to temperatures of ca. 350°C at the inlet, and this component should not be insulated (see Cyclones).

As was mentioned with carbon monoxide, because the system is under pressure, any small explosions caused by air leaks will also propagate inwards. That is not to encourage complacency; air leaks, particularly at the reactor top or where the gases are dirty and hot, e.g. around the pyrolysis zone or prior to the cyclone, are the danger zones for localised explosion as they can lead to high temperature and damage to the component structural integrity. It is therefore good practice to complete regular leak tests. The best way to do this - rather than pressurising the reactor and using soapy water or smoke release methods – is to use the flare stack bypass blowers, but there will need to be some internal pressure monitor (manometer). This can be achieved before igniting the gasifier, so while the system is cold. Turn on the fans and block the air inlet so that the system builds up pressure. Then, while still covering the air inlets, shut off the fan and see that pressure does not drop instantly which would indicate a gross leak; although very gradual decrease in pressure is expected.

As explained, a gasifier is not a pressure vessel. Still, for closed-topped systems, a simple pressure release valve set at low pressure is often incorporated as a belt and braces safety feature. They are cheap and effective.

The flame from flared gas is controlled and not likely to contact any external combustible material (assuming that there is adequate clearance above the flare stack outlet). The main risk of explosion comes from hydrogen and by reckless operators opening the reactor while it is still hot. People have misconceptions about the dangers of hydrogen, probably due in part to the Hindenburg airship disaster. It is very true that hydrogen has an extremely wide flammability range in air (between 4% and 75%) meaning that it can easily ignite when it is released, and it also burns with an invisible flame. But, I have met with safety experts who say that as a fuel it is much safer than petrol for many reasons: its lower explosive limit (13%) is higher than its lower flammability limit so H_2 generally burns rather than explodes (62). Hydrogen is also very light, and any escapes will rapidly dissipate (62). Hence, as long as working in the open air, the potential for harm is greatly reduced. As a final thought on this, consider that oil and petrol are highly flammable, carcinogenic liquids, which are widespread, globally transported, stored and distributed with relative safety. Producer gas just poses a different type of hazard.

Occasional explosions at the engine air/fuel intake (sounding like backfiring) are not uncommon. These occur more often when the engine runs slowly on high load, and they are nothing to worry about.

A further fire risk is with the char, particularly the small pieces in the ash grate, if somehow it comes in contact with an ignition source. It is therefore advisable to wait until the reactor has completely cooled before emptying the ash grate.

10.4. environmental: air

In comparison to direct burning of wood for domestic cooking, gasification of small-scale biomass emits much fewer airborne pollutants under general operation. This is an important benefit as air pollution is a big killer – approximately 7 million premature deaths in 2012 (63).

Because there are much lower concentrations of sulphur in biomass than in coal (31, 64), the atmospheric pollutant SO_x is therefore at near zero with biomass gasification and engine combustion (30). Neither SO_x nor the other gaseous pollutant NO_x will initially occur to the level that they would otherwise do in combustion/incineration as a gasifier operates with limited oxygen. However, since the ultimate fate of the producer gas will likely be combustion and assuming that air (with 79% N_2 content) is used throughout, NO_x emissions will occur as they would from any combustion exhaust. Yet, NO_x emissions are reportedly much lower from producer gas engines likely due to the way that the engine is adjusted to cater for this type of fuel (30, 65).

There have been very few studies that have assessed the level of particulates from an engine exhaust when fuelled by producer gas. They are also predicted to be low in comparison to diesel and petrol engines (30); and this has been verified in one series of tests (65).

The likelihood of dioxin production in a gasifier is less certain, but some research suggests that it is low, and there is evidence to support this. Firstly, tests found that during pyrolysis and gasification 30 to 60% of the chlorine content in biomass remained in the char, and furthermore dioxin precursor formation was inhibited by the process inherently having a low availability of O_2 (46).

Chlorine concentrations increase in shoots, leaves, and stems, and grassy species, particularly where these plants grow near to sea water (46); but, wood has a much lower chlorine content, therefore reducing the possibility of dioxin release. Heartwood contains chlorine in concentrations of between 0.005 to 0.009 % with bark and twigs between 0.009 to 0.03% (66). Coal has higher chlorine concentrations of 0.07 – 0.7% (67). The situation is very different with Municipal Solid Waste (MSW), which has chlorine in concentrations of about 1 wt% originating in plastics and common salts (61), hence dioxin release from MSW incineration and other thermal conversion technologies is a matter of concern. This is discussed more fully in later sections.

Small solid particles cause chronic respiratory illness, and the operator is exposed to these when cleaning out the reactor and sieving wood chip. Particulates larger than 10 microns are removed in the nose and throat. Those between 5 to 10 microns (PM_{10} = particulates smaller than 10 microns) are captured in trachea and bronchi and can cause lung problems. But the smaller particulates ($PM_{2.5}$ = particulates less than 2.5 microns) are able to pass into the bloodstream and are of most concern for health (68). Cyclones will capture the larger particulates and these are safely collected in its bottom collection trap, and the finer polishing filters will capture most of the rest. Care must then be taken to avoid inhalation when emptying these filters and when cleaning out the char from the ash grate.

10.5. environmental: water

The toxicity of tar to the aquatic environment when dissolved in water has already been mentioned, and it is likely that there will be local legislation in place which prohibits its disposal into drains and water courses. If the gasifier has a hopper condensate collector, then the dissolved hydrocarbons in this condensate are of sufficient dilution to be dealt with by on-site separation and filtering through wood screenings or straw. This can then be burnt. Tar collected from the post reactor gas however is different. In liquid form it is more concentrated and it will likely contain higher class PAHs. In theory this could also be poured on feedstock and re-introduced, with sufficient care for operator safety of course. That said, the volumes are minimal so disposal costs will not be high, unless however wet scrubbing systems are used in which can the volumes of spent scrubbing water can be significant.

11. other considerations

11.1. vehicular gasifiers

Nearing the end of the book, it has not yet been necessary to discuss whether the small gasifier is to be a stationary or vehicular-mounted system. The principles of operation are exactly the same. Many of the reference sources cited so far, and listed again in the Bibliography, discuss in detail biomass gasifiers attached to vehicles (cars, lorries, tractors, and boats). A vehicular gasifier has a number of unique, practical issues that have relevance. Firstly, the weight of the total system will be an extra consideration which puts limitations on the gas cleaning components. Adding to this will be the amount of wood feedstock that is needed for at least one round trip.

Performance of a producer gas vehicle in one respect will not be comparable to one using petrol or diesel fuel, as there will likely be weak power when accelerating from a road junction and uphill. This is just something to get used to in comparison to manufactured vehicles that are made to perform (and "sell") rather than be economical or efficient. Straight line speed and long distance driving performance should be no different.

With vehicular gasifiers, vibration helps to overcome bridging problems in the feedstock hopper, and the cooling effect of transit can be used to enhance the efficiency of a finned-tube gas cooler/condenser. There will be greater restrictions though in what is possible for the routing of pipework, and some reactor controls and displays will need to be observed and manipulated from the driver's dashboard, such as thermocouple readings and a manual carburettor choke.

It is also important to note that in some countries, operating a gasifier vehicle might not be legal. This is presently true for example in Britain with only liquid petroleum gas (butane or propane) fuelled vehicles permitted by law *on the public road* (69). This does not preclude using gasifier vehicles off road however.

11.2. stationary operation: legal and financial

It is unlikely that there will be any legal restrictions on the purchase and operation of a stationary gasifier. In many large urban areas, there can be legislation for smoke abatement. Smoke however does not come from a gasifier at steady state operation, so systems will be able to operate legally even in smoke control areas, particularly since this legislation usually pemits smoke at start up as with wood fires. Start-up is the only time that gasifiers produce smoke (70).

For static systems, off-grid operation is simple and will give energy independence. Connection can however still be made to the electricity distribution network but some synchronisation unit will be needed and there is the likelihood of a fee payable to the network operator to ensure compatability. This may then provide extra financial incentives if subsidies are available for renewable electricity exportation. You may be lucky to find that incentives are also available for system purchase as well as operational outputs. These incentives vary from country to country, and they are also permanently changing, so no attempt is made here to adequately address this subject as the information becomes rapidly obsolete following publication.

11.3. biochar

When explorers visited the Amazon rainforest in the 1870s they discovered numerous patches of extremely rich soil isolated within regions where fertility was otherwise poor (71). These soils became known as the *terra preta do indio*, and the pioneering work to investigate them was undertaken by Wim Sombroek in the 1950s and 1960s (72). Others have continued his work since then.

The *terra preta do indio* were found to be extremely long lasting (up to 7000 years old). They contained three times more nitrogen and phosphorous than surrounding land, and seventy times more carbon. Most importantly, the principle ingredient was charcoal (71). Modern terminology has now applied the definition that when charcoal is applied to soil it is "biochar", but the substance is still the same material that is obtained from pyrolysis, gasification, or forest fires.

Since the 1960s much more has been written about the potential of biochar application. Barrow lists 26 benefits, such as enhancing plant growth, reducing greenhouse gas emissions by suppressing CH_4 and N_2O release, increasing water and nutrient retention, countering land degradation/land reclamation, balancing pH, replacing the need for petrochemical fertilisers, encouraging beneficial soil micro-organisms, and sequestering atmospheric carbon (73). There is still a big mystery however, because no-one knows how the *terra preta do indio* was made, nor therefore how to recreate it.

High crop yields do not just come by applying charcoal to soil (74), so the *terra preta do indio* was not a product of conventional shifting cultivation "slash and burn" activity (75). Tests have been made to determine whether, during the char production process, the beneficial properties can be recreated by changes in temperature, residence time, initial feedstock type, and heating rate; but results have been inconsistent or inconclusive (74). Even where initial fertility improvements have been noticed, no long-lasting effect has been reproduced (75). There is something else still to be discovered. The *terra preta do indio* are known to have been formed by human activity (either intentionally for agriculture, or as a byproduct of land occupation). Suggestions about their origin range from the burning of green vegetation under rainfall, to the inclusion of numerous ingredients such human and animal excrement (part of ancient midden heaps), algae, local vegetable or aquatic animal oils (73, 74).

The Amazonians will not have operated downdraft gasifier to engine systems, however in previous chapters it has been described how reactor conditions affect the variability of gas, tar and soot. This is interesting because char from a gasifier is particularly rich in polycyclic aromatic hydrocarbons and it is believed that the high amounts of these compounds make the carbon in the *terra preta* last so long in the soil. The theory is that with time these molecules become partially oxidised generating edges which have a high nutrient bonding capacity (74).

Because some PAHs are carcinogenic, there is a concern that applying biochar to soil could pose a risk to the food chain. It appears however that the PAHs are very resistant to leaving char; in fact they remain over geological time periods. Charcoal from forest fire deposits in Triassic, and Permo-Triassic boundary layers have PAH contents in the same range as those from recently produced pyrolysis (76). In addition to this, there is controversy as to whether the PAHs actually become bio-available or not.

Tests to assess whether PAHs in charcoal can affect plant growth have provided an equal number of positive to negative effects (77, 78). To also put this into context, urban and forest soils, cattle and pig manure have PAH concentrations in the range similar to those of charcoal (74, 79). Unsurprisingly therefore, only a few countries have attempted to offer suggestions for legal limits of PAH in char when used as a soil amendment. The recent European Biochar Guidelines (80) have taken one of these and made a first attempt at suggesting threshold values. They define two grades of biochar and in these, with respect to a selected number of 16 PAH molecules, they propose concentrations of 12 mg.kg^{-1} (basic grade char) and 4 mg.kg^{-1} (premium grade). The scientific basis for this is not strong, and the guidelines admit that the risk of PAH contamination is "...considered to be low, even if higher thresholds would be taken into account".

So, using gasifier char on soil could be enormously beneficial. It contains all the same mineral elements as wood ash, plus maybe something more. However, there is also the possibility of some risk. Much more information is available for the interested reader, but at present the truth is still undiscovered.

12. future

For someone to have got this far, it should I hope occur to them that gasifiers have many attractions and even greater future potential; but also that there are difficulties with widespread adoption. The present technology is rigid in respect to variations in feedstock, and society is attuned to machines that need less preparatory care. To a lesser extent, although the feedstock resource is available locally to supply a gasifier market, the current infrastructure of wood harvesting and supply needs some re-adjustment, much as it was when vehicular gasification was deployed approximately eighty years ago. Equally, perceptions of gasification need to be improved. This book has attempted to address this, but the real need is for modern demonstration sites. From this it would be sensible to hope that a more holistic and integrated approach to community waste and energy would develop with small gasifier systems providing an incentive for local and communal bio-economies built around the sustainable management of land.

These obstacles to biomass gasification do not seem at all great. The technological challenges of making gasifiers more generic in response to feedstock and with labour-saving controls seem realistically achievable given sufficient research focus and the current state of engineering knowledge that has been applied successfully elsewhere. The driver for this is already present with the unavoidable truth that the easy availability of energy comes at too big a price to last.

It has been made clear that moisture aside, the chemical composition of any potential feedstock is not the main criteria to stop gasifiers being generic waste-to-energy conversion systems, more its physical properties. With seemingly little difficulty, any shredded material can be shaped and compacted into pellets or briquettes, and to manufacture these reconstituted materials against friability, thermal decomposition, or steamed disintegration.

Pellets and briquettes can provide homogeneity, easier handling, and optimum internal flow characteristics, in addition to providing a means to utilise lots of waste biomass material - particularly dusty, fibrous, and other non-woody substances. Even previously reconstituted waste wood such as MDF and chipboard should not be a problem compositionally. The formaldehyde glues used in chipboard, etc will burn. It is the physical rigidity that is questionable as they are reconstituted and so prone to fall apart under thermal and attritional stress. To make the pellets as good as possible, they should be larger than those currently manufactured for biomass boilers/room heaters and compressed very tight. If through some research effort they could be pre-treated to make them resistant to decay then it can be envisaged that this would open the door to small-scale gasifier uptake in a similar way to that which has occurred with the small wood pellet appliances.

One way of strengthening reconstituted wood pellets against disintegration is by a process called torrefaction. This involves gentle roasting, up to about 200°C, as is done with coffee beans. The torrefaction process does indeed improve the durability of reconstituted or non-woody bio waste for gasification applications (81, 82). The reasons why can be seen in Figure 36 which shows the torrefied pellets in a flask in comparison to the virgin wood pellet (Miscanthus in this case). 100 ml of room temperature tap water was added to each flask containing 200 ml of each pellet type. After ten minutes the Miscanthus pellet had absorbed all the water and had swollen, completely losing all its original form. The torrefied pellet did not absorb any water and when it was removed after two days, it had not changed at all.

I trialled these torrefied pellets in the Power Pallet downdraft gasifier. They seemed excellent and had high bulk density, and a clear blue flame occurred almost instantly. However, the auger feeder crushed them, and they were also "too good" for the gasifier as the internal temperature rapidly became too hot, which evidences the fact that gasifiers are currently designed for a limited type of feedstock only (in this case, one with a certain percentage of moisture). More work is needed with trials on both manufacture and gasification of torrefied pellets. On the reactor side, this example highlights that the future of small-scale gasification will be the design of a system that can accommodate feedstock variability by perhaps having interchangeable inserts or smarter control.

Fig. 36: water absorbance of commercial wood pellets in comparison to torrefied pellets after contact with 100 ml of cold tap water.

So far this book has been considering a woody feedstock with relatively uniform volatile to fixed carbon ratios, total ash content of much less than 1% by dry weight, and chlorine contents of at least ten times less than this. Gasification of MSW has been touched upon, and it is this which can really have a big impact on society because of the enormous and problematic volumes of mixed waste from cheap plastics and expanding consumerism - global MSW tonnage is expected to be 2.9 billion tonnes per year in 2022, - a 45% increase from 2011 levels (83).

Many industrial waste-to-energy gasification plants have been built and many more are being proposed. They are appealing to investors partly because incineration has associated negative publicity. There is also the positive attraction that gasification has not been subjected to modern development like most other technologies, and as such it has something of a "sleeping giant/waiting to happen" aspect to it. This makes investors interested in "getting in on the ground". Through "green" investment schemes and subsidies, governments are helping to drive this forward too. But gasifiers are not yet at this stage of development, and the reader should by now understand why. Firstly there is the inherent problem of heterogeneity in both physical and chemical properties which means that extensive sorting and pre-treatment is essential (84). Ash content in MSW can also be high at as much as 25% (31) of which there is a high risk of toxic metals content. Chlorine is also much higher in MSW (61) and this is serious because it forms HCl, and also because it is a precursor for dioxins (46). For these reasons, problems continue to beset attempts at mixed waste gasification, such that failures are commonplace and many countries have completely stopped all attempts at trying. One way that plant designers get around this is to have two stage or close coupled reactors in which greater quantities of oxygen are added. This has led to accusations that gasification of MSW is really just "incineration in disguise" (85).

A different but very interesting concept is to make use of gasifier engine exhaust for enriching the air in greenhouses. This exhaust gas is a higher quality CO_2 composition and so it could be used to create new carbon sinks (86). The application would also permit a way to bypass the problems of tar as the gas could be burned in a boiler rather than an engine.

Producer gas can in theory be supplied to some types of fuel cells, which are more efficient than the internal combustion engine, but this would then require that the gas is much cleaner of tar. Unless cheap and sustainable ways of making and storing hydrogen fuel appear (which at present seems highly unlikely), and if society wants to keep the internal combustion engine rather than go back to draught animals, then small-scale biomass gasifiers would have to be returning very soon just as they did over the last century. Either through choice or by enforcement, we must accept that *"there is no such thing as a free meal"* (87).

13. bibliography

13.1. recommended reading

(1) Kaupp, A. State of the art for small-scale gas producer-engine systems, (1984), First published by the German Appropriate Technology Exchange (GATE), Reissued by the Biomass Energy Foundation Press: Golden. *This book gives a good background and history. Definitely one of a handful of significant works. Chapters 3 and 4 are good but now a little dated in their content as scientific understanding has moved on somewhat. Replicated data from the Swedish experiences.*

(25) LaFontaine, H., Zimmerman, F.P. Construction of a simplified wood gas generator for fueling internal combustion engines in a petroleum emergency, (1989), Federal Emergency Management Agency: Washington. *Written by a former circus owner, who also founded the Biomass Energy Foundation in USA after previously manufacturing gasifiers during World War Two while working for the Danish underground. As might be expected, this is a bit of an oddity. It describes how to build a gasifier from a metal rubbish bin if there is an emergency so that it can be used to power a vehicle. However, it is written based on obvious experience and a great depth of knowledge.*

(26) Food and Agriculture Organization. Woodgas as an engine fuel, Forestry Paper 72, (1986), United Nations: Rome. *Very well written with a practical focus, and also available as an online resource. Contains sizings and good background information.*

(27) Generator Gas – The Swedish Experience from 1939 – 1945, The Swedish Academy of Engineering Sciences, SERI/SP-33-140, (1979), Solar Energy Research Institute: Colorado. *Devoted to charcoal gasification as well as biomass. Summarises the technical, scientific and commercial information from the Second World War when Sweden converted 40% of its vehicles to wood gas power. First*

published in 1959. Very detailed explanations of history, specifics of deigns, performance and operational aspects.

(31) Reed, T., Das, A. Handbook of Biomass Downdraft Gasifier Engine Systems, (1988), Solar Energy Research Institute: Colorado.

(32) Skov, N.A., Papworth, M.I. Driving on Wood. The lost art of driving without gasoline. Revised 2006 with plans for building a WW2 gasifier, (2006), The Biomass Energy Foundation Press: Franktown.

(33) Jain, B.C. Commercialising biomass gasifiers: Indian experience, (2000), *Energy for sustainable development*, **4** (3), pp.72-82. *A good outline of the most recent prolonged attempt at commercialising small-scale bioresource gasifiers. What worked, what didn't, and the same issues repeated in works by Kaupp. Described efforts focussed on field testing and user preferences.*

(35) Rabou, L.P.L.M., Zwart, R.W.R., Vreugdenhil, B.J., Bos, L. Tar in Biomass Producer Gas, the Energy Research Centre of the Netherlands (ECN) Experience: An Enduring Challenge, (2009), *Energy and Fuels*, **23**, pp. 6189-6198.

(40) Kaupp, A. Gasification of Rice Hulls theory and praxis, (1984). Deutsche Gesellschaft fur Technische Zusammenarbeit (GTZ) GmbH: Eschborn. *Another excellent book by Albrecht Kaupp. Focuses on rice hull gasification which involves fine high silica-rich material, but the book is also strong on theory and praxis of small gasifiers in general.*

(44) Mukunda, H.S., Dasappa, S., Paul, P.J., Rajan, N.K.S. Biomass to Energy, (2003), ABETS Indian Institute of Science: Bangalore. Available online at: http://cgpl.iisc.ernet.in

(49) Milne, T.A., Evans, R.J. Biomass Gasifier "Tars": their nature, formation, and conversion – NREL/TP-570-25357, (1998), National Renewable Energy Laboratory: Golden.

(56) Stassen, H.E. Small-scale biomass gasifiers for heat and power: a global review. World bank technical paper no 296, (1995), The World Bank: Washington DC.

13.1. recommended websites/resources

http://www.driveonwood.com/ *Most of the recommended texts are available from this website which also has lots of other material split into topic headings along with sections giving advice, and a members forum.*

http://www.thersites.nl/ *Excellent website on tars and how they form from ECN.*

http://www.chemguide.co.uk/ *No information about gasification, but this is an accompanying website for the excellent chemistry book: Clark, J. Calculations in AS/A Level Chemistry, Harlow: Pearson.*

13.2. references

(1) See Recommended reading.

(2) H.M. Government. UK Renewable Energy Road Map 11D/698, (2011), Department of Energy and Climate Change: London.

(3) H.M. Government. UK Bioenergy Strategy, (2012), Department of Energy and Climate Change: London.

(4) Anseeuw, W., L. Alden Wily, L. Cotula, and M. Taylor. Land rights and the rush for land: findings of the global commercial pressures on land research project, 2012, ILC: Rome.

(5) CAFOD. Enough food for everyone IF, Policy Report, (2013), Christian Aid Communications Division: London.

(6) Deininger, K., and Byerlee, D. Rising Global Interest in Farmland: Can it yield sustainable and equitable benefits?, (2011), World Bank: Washington.

(7) Righelato, R., Spracklen, D.V. Carbon mitigation by biofuels or by saving and restoring forests, Science, 2007, 317 (5480), p. 902.

(8) The World Bank. Electric power transmission and distribution losses (% of output) 2012 [online], (2015), accessed 18th August 2015. Available from: http://data.worldbank.org/indicator/EG.ELC.LOSS.ZS

(9) U.K. Parliament, Department of Energy and Climate Change. Electricity: Chapter 5, Digest of United Kingdom Energy Statistics [online], (2015) accessed 22nd September 2015. Available from: https://www.gov.uk/government/statistics/electricity-chapter-5-digest-of-united-kingdom-energy-statistics-dukes

(10) International Energy Agency Statistics, Indicators for 2012 [online], (2015), accessed 18th August 2015. Available from: http://www.iea.org/statistics/statisticssearch/report/?&country=UK&year=2012&product=Indicators

(11) U.K. Parliament, Department of Energy and Climate Change. Final UK greenhouse gas emissions national statistics: 1990-2013 [online], (2015), accessed 18th August 2015. Available from: https://www.gov.uk/government/statistics/final-uk-emissions-estimates

(12) OFGEM Annual Sustainability Report Dataset, (2012), accessed 11th April 2014. Available from: https://www.ofgem.gov.uk/publications-and-updates/annual-sustainability-report-2011-2012.

(13) Whittaker, C. The nature of the wood pellet supply to the UK, (2014), In: Torrefaction Workshop, Leeds, 2-3 April 2014.

(14) Forest Research. Woodchip Drying Project Report, (2011), Forestry Commission: Delamere.

(15) McGovern, R.E. Report on trials of farm grain dryers for woodchip moisture

Reduction, (2007), SAC Consultancy Services Division report: Aberdeen.

(16) Kreschmer, B., Watkins, E., Baldock, D., Allen, B., Keenleyside, C., Tucker, G. Securing biomass for energy – developing an environmentally responsible industry for the UK now and into the future, (2011), Institute for European Environmental Policy: London

(17) Bowyer, C., Baldock, D., Kretschmer, B., Polakova, J. The GHG emissions intensity of bioenergy: does bioenergy have a role to play in reducing GHG emissions of Europe's economy?, (2012), Institute for European Environmental Policy: London.

(18) Gadde, B., Bonnet, S., Menke, C., Garivait, S. Air pollutant emissions from rice straw open field burning in India, Thailand and the Philippines, (2009), *Environmental Pollution*, **157**, pp. 1554-1558.

(19) ISAT/GTZ. Biogas Digest Volume 1. Biogas Basics – Information and advisory service on appropriate technology (ISAT), Deutsche Gesellschaft für Technische Zusammenarbeit (GTZ); 1999.

(20) Forestry Commission. A woodfuel strategy for England, (2007), Forestry Commission Publications: Stockport.

(21) Bradford, D. Biomass applications utilised by Barnsley Metropolitan Council, (2006), *International Journal of Low Carbon Technologies*, **1** (4), pp.343-354.

(22) House of Lords. Waste or resource? Stimulating a bioeconomy, House of Lords Science and Technology Committee, 3rdReport of Session 2013-2014, (2014), H.M. Stationary Office: London.

(23) DEFRA. Wood waste landfill restrictions in England call for evidence, (2012), DEFRA: London.

(24) Antal Jr, M.J., Grønli, M. The art, science, and technology of charcoal production, (2003), *Industrial and Engineering Chemistry Research*, **42**, pp.1619-1640.

(25) See Recommended reading

(26) See Recommended reading

(27) See Recommended reading

(28) Horsfield, B.C. History and potential of air gasification. In: Reftofit 79' Proceedings of a Workshop on Air Gasification, Seattle, Washington, 2nd February 1979, SERI/TP-49-183, (1979), The Solar Energy Research Institute: Colorado.

(29) Goss, L.R. The downdraft gasifier. In: Reftofit 79' Proceedings of a Workshop on Air Gasification, Seattle, Washington, 2nd February 1979, SERI/TP-49-183, (1979), The Solar Energy Research Institute: Colorado.

(30) Sridhar, G., Sridhar, H.V., Dasappa, S., Paul, P.J., Rajan, N.K.S., Mukunda, H.S. Development of producer gas engines, 2005, *Proceedings of the Institution of Mechanical Engineers, Part D: Journal of Automotive Engineering*, **219** (3) pp. 423-438.

(31) See Recommended reading

(32) See Recommended reading

(33) See Recommended reading

(34) ALL Power Labs. Personal scale power [online], (2015), accessed 16th September 2015. Available from: http://www.allpowerlabs.com/

(35) See Recommended reading

(36) Denton, M.J. The place of life and man in Nature: defending the anthropocentric thesis, (2013), *Bio-complexity*, **1** (1), pp. 1-15.

(37) Okumura, Y., Hanaoka, T., Sakanishi, K. Effect of pyrolysis conditions on gasification reactivity of woody biomass-derived char, (2009), *Proceedings of the Combustion Institute,* **32**, pp. 2013-2020.

(38) Zainal, Z.A., Ali, R., Lean, C.H., Seetharamu, K.N. Prediction of performance of a downdraft gasifier using equilibrium modelling for different biomass materials (2001), *Energy conversion and management*, **42**, pp. 1499-1515.

(39) Jirjis, R., Pari, L., Sissot, F. Storage of poplar wood chips in northern Italy, (2008), *World Bioenergy*, **8**, pp. 85-89.

(40) See Recommended reading

(41) Mayerhofer, M., Mitsakis, P., Meng, X., de Jong, w., Spliethoff, H., Gaderer, M. Influence of pressure, temperature and steam on tar and gas in allothermal fluidized bed gasification, 2012, *Fuel*, **99**, pp. 204-209.

(42) Huang, S., Wu, S., Wu, Y., Gao, J. The physiochemical properties and catalytic characteristics of different biomass ashes, (2014), *Energy Sources, Part A: Recovery, Utilization and Environmental Effects*, **36** (4), pp. 402-410.

(43) Rollinson, A.N., Karmakar, M.K. On the reactivity of various biomass species with CO_2 using a standardised methodology for fixed-bed gasification, (2015), *Chemical Engineering Science*, **128**, pp. 82-91.

(44) See Recommended reading

(45) Middleton, F.A., Bruce, C.S. Engine tests with producer gas, (1946), *Journal of Research of the National Bureau of Standards*, **36**, pp. 171-183.

(46) Björkman, E., Strömberg, B. Release of chlorine from biomass at pyrolysis and gasification conditions, (1997), *Energy and Fuels*, **11**, pp. 1026-1032.

(47) Siplä, K., Kuoppala, E., Fagernäs, L., Oasmaa, A. Characterization of biomass-based flash pyrolysis oils, (1998), *Biomass and Bioenergy*, **14** (2), pp. 103-113.

(48) Groeneveld, M.J. The co-current moving bed gasifier, Ph.D thesis, (1980), Twente University of Technology, The Netherlands.

(49) See Recommended reading

(50) Vreugdenhil, B.J., Zwart, R.W.R., Tar formation in pyrolysis and gasification, ECN-E—08-087. 2009, Energy Research Centre of the Netherlands.

(51) Kiel, J.H.A., van Paasen, S.V.B., Neeft, J.P.A., Devi, L., Ptasinski, K.J., Janssen, F.J.J.G, Meijer, R., Berends, R.H., Temmink, H.M.G., Brem, G., Padban, N., Bramer, E.A. Primary measures to reduce tar formation in fluidised-bed biomass gasifiers. Final report SDE project P1999-012, ECN-C—04-014. (2004), Energy Research Centre of the Netherlands.

(52) Biomass gasification – Tar and particulates in product gases – sampling and analysis, British Standard DD CEN/TS 15439:2006, (2006), BSI: London.

(53) Van Paasen, S.V.B., Kiel, J.H.A. Tar formation in a fluidised bed gasifier – impact of fuel properties and operating conditions, ECN-C—04-013, (2004), Energy Research Centre of the Netherlands.

(54) Donnot, A., Magne, P., Deglise, X. Flash Pyrolysis of Tar from the Pyrolysis of Pine Bark, (1985), *Journal of Analytical and Applied Pyrolysis,* **8**, pp. 401–414.

(55) Parikh, P.P., Bhave, A.G., Kapse, D.V., Ketkar, A., Bhagivat, A.P. Design and Development of a Wood Gasifier for I.C. Engine Applications - New Approach for Minimization of Tar, (1988), in Research in Thermochemical Biomass Conversion. Edited by A.V. Bridgwater and J.L. Kuester, Elsevier Applied Science: London, pp. 1071–1087.

(56) See Recommended reading

(57) Phuphuakrat, T., Namioka, T., Yoshikawa, K. Absorptive removal of biomass tar using water and oily materials, (2011), *Bioresource Technology*, **102**, pp. 543-549.

(58) Shackley, S., Carter, S., Knowles, T., Middelink, E., Haefele, S., Sohi, S., Cross, A., Haszeldine, S. Sustainable gasification-biochar systems? A case study of rice-husk gasification in Cambodia, Part 1: Context, chemical properties, environmental and health and safety issues, 2012, Energy Policy, 42, pp. 49-58.

(59) Her Majesty's Inspectorate of Pollution, 1994. Environmental Protection Act 1990, Technical Guidance Note A3 (Abatement) Pollution abatement technology for particulate and trace gas removal, HMSO: London.

(60) Zwart, R.W.R., Van der Drift, A., Bos, A., Visser, H.J.M., Cieplik, M.K., Könemann, H.W.J. Oil-based gas washing – the flexible tar removal for high efficient production of clean heat and power as well as sustainable fuels and chemicals, *Environmental Progress and Sustainable Energy*, 28 (2009) pp324-334.

(61) Nordgreen, T. Iron-based materials as tar cracking catalyst in waste gasification, 2001, PhD thesis, KTH – Royal Institute of Technology, Sweden.

(62) Rand, D.A.J., Dell, R.M. Hydrogen energy, challenges and prospects. Cambridge: RSC Publishing, 2008, pp. 146-178.

(63) World Health Organisation, 7 million premature deaths annually linked to air pollution [online], (2014), accessed 7th September 2015. Available from: http://www.who.int/mediacentre/news/releases/2014/air-pollution/en/

(64) Singh, S., Ram, L.C., Masto, R.E., Verma, S.K. A comparative evaluation of minerals and trace elements in the ashes from lignite, coal refuse, and biomass fired power

plants, (2011), *International Journal of Coal Geology*, **87**, pp.112-120.

(65) Hamilton, J.E., Adams, J.M., Northrop, W.F. 2014. Particulate and aromatic hydrocarbon emissions from a small-scale biomasss gasifier-generator system, *Energy and Fuel*. **28** (5), pp. 3255-3261.

(66) Werkelin, J., Skrifvars, B-J., Zevenhoven, M., Holmbom, B., Hupa, M. Chemical forms of ash-forming elements in woody biomass fuels, (2010), *Fuel*, **89**, pp. 481-493.

(67) Huggins, F.E., Huffman, G.P. Chlorine in coal: an XAFS spectroscopic investigation, (1995) *Fuel*, **74** (4), pp. 556-569.

(68) World Health Organisation, Fact Sheet 313, Ambient (outdoor) air quality and health, (2014), World Health Organisation: Geneva.

(69) Great Britain Parliament. The Road Vehicles (Construction and Use) Regulations 1986. No. 1078, Part IV, D, Regulation 94.

(70) Great Britain Parliament, Clean Air Act 1993.

(71) Marris, E. Black is the new green, (2006), *Nature*, **442**, pp. 624-626.

(72) Sombroek, W.G. Amazon Soils. A Reconnaissance of the soils of the Brazilian Amazon region, (1966), Centre for Agricultural Publications and Documentation (PUDOC): Wageningen.

(73) Barrow, C.J. Biochar: potential for countering land degradation and for improving agriculture, (2012), Applied Geography, 34, pp. 21-28.

(74) Schimmelpfennig, S., Glaser, B. One step forward toward characterization: some important material properties to distinguish biochars, (2012), *Journal of Environmental Quality*, **41**, pp. 1001-1013.

(75) Glaser, B. Prehistorically modified soils of central Amazonia: a model for sustainable agriculture in the twenty-first century, (2007), *Philosophical Transactions of the Royal Society*, **362**, pp. 187-196.

(76) Bucheli, T.D., Hilber, I., Schmidt, H-P. Polycyclic aromatic hydrocarbons and polychlorinated aromatic compounds in biochar, (2015), in: Lehman, J., Joseph, S. Biochar for Environmental Management, science, technology and implementation (2nd Ed.), Earthscan: London.

(77) Rogovska, N., Laird, D., Cruse, R.M., Trabue, S., Heaton, E. Germination tests for assessing biochar quality, (2012), *Journal of Environmental Quality*, **41**, pp. 1014-1022.

(78) Koltowski, M., Oleszczuk, P. Toxicity of biochars after polycyclic aromatic hydrocarbons removal by thermal treatment, (2015), *Ecological Engineering*, **75**, pp. 79-85.

(79) Hale, S.E., Lehmann, J., Rutherford, D., Zimmerman, A.R., Buchmann, R.T., Shitumbanuma, V., O'Toole, A., Sundqvist, K.L., Arp, H.P.H., Cornelissen, G. Quantifying the total and bioavailable polyclclic aromatic hydrocarbons and dioxins in biochars, (2012), *Environmental Science and Technology*, **46**, pp. 2830-2838.

(80) EBC. European Biochar Certificate – guidelines for a sustainable production of biochar, version 6.1 of 19th June 2015, (2015), European Biochar Foundation: Arbaz.

(81) Deng, J., Wang, G-J, Kuang, J-H., Zhang, Y-L, Lou, Y-O. Pretreatment of agricultural residues for co-gasification via torrefaction, (2009), *Journal of Analytical and Applied Pyrolysis*, **86**, pp. 331-337.

(82) Prins, M.J., Ptasinski, K.J., Janssen, F.J.J.G. From coal to biomass gasification: Comparison of thermodynamic efficiency, (2007), *Energy*, **32**, pp. 1248-1259.

(83) Lusardi, M.R., McKenzie, K., Themelis, N.J., Castaldi, M.J. Technical assessment of the CLEERGAS moving grate-based

process for energy generation from municipal solid waste, (2014), *Waste Management and Research*, **32** (8), pp. 772-781.

(84) Rollinson, A.N. Re: EPR/RP3132ND/A001 CPP Washwood Heath online], (2015). Available from: http://ukwin.org.uk/files/pdf/Rollinson_March_2015_Washwood_Heath_Assessment_EPR_RP3132ND_A001.pdf

(85) Global Alliance for Incinerator Alternatives and Greenaction for Health and Environmental Justice. Incinerators in Disguise, (2006). Available from: http://www.no-burn.org/article.php?id=283

(86) Dion L-M, Lefsrud M, Orsat V, Cimon C. Biomass gasification and syngas combustion for greenhouse CO_2 enrichment, (2013), *Bioresources* **8**, pp. 1520–1538.

(87) Commoner, B. The Closing Circle: confronting the environmental crisis, (1972), Cape: London.

appendix A: reactor sizings

| Dimensions | | | | | | | | Gas output range (Nm³.hr⁻¹) | | Maximum wood consumption [f] | Air velocity |
dh (mm)	dr (mm)	d'r (mm)	h (mm)	H (mm)	R (mm)	A	dm (mm)	Max	Min	kg.hr⁻¹	m.s⁻¹
60	268	150	80	256	100	5	7.5	30	4	14	22.4
80	268	176	95	256	100	5	9	44	5	21	23.0
100	268	202	100	256	100	5	10.5	63	8	30	24.2
120	268	216	110	256	100	5	12	90	12	42	26.0
100	300	208	100	275	115	5	10.5	77	10	36	29.4
115	300	228	105	275	115	5	11.5	95	12	45	30.3
130	300	248	110	275	115	5	12.5	115	15	55	31.5
150	300	258	120	275	115	5	14	140	18	67	30.0

Sizings for the combustion zone of a downdraft gasifier reactor. Extracted from (1). For other dimensions see (25, 26, 27, 32). See over for key.

NB: The table shows the gasifier operating at maximum capacity under a full draw of power. The average wood usage will likely be half this.

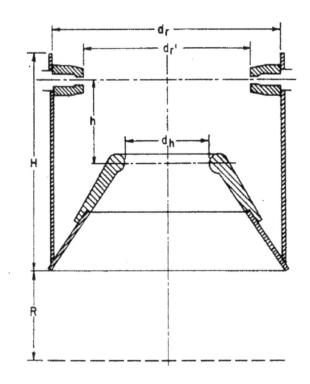

appendix B: operational schedule

As the right conditions can be achieved by precise internal configuration, by following certain rules, successful gasifier operation can be achieved. The following routine tasks can be adjusted based on observations as experience develops.

daily, before each start up

- Ensure that there is sufficient feedstock with exactly the right size range and moisture content.
- Empty soot from the cyclone hopper, char/ash from below the reactor grate, and condensate from the reactor or filter condensate traps.
- Check the integrity of all connectors, lids and seals. Then do a leak test using the fan blowers.

fortnightly (ca. 50 hours of operation)

- Clean the blower fans of tar using methylated spirit. Flush any coolers with water.

ongoing

- Put effort into getting high quality feedstock of the right size, with as few fines as possible. If using a dry wood filter or other packed bed material, this will need changing about once per month, if the gasifier has been operated properly (and not had dirty or moist gas flowing excessively through the filters).
- Oiled foam or fabric filters will likely need cleaning, flushing or changing.

- General engine care will be required, such as topping up oil and coolant, cleaning spark plugs, etc.
- Regularly check and clean the engine fuel intake/throttle valve for soot and tar build-up.

important

It has been stressed throughout this book that the operator is one of the three primary components for successful gasifier operation. If things begin to go wrong, do not overlook whether the operator was perhaps a little bit complacent checking system components (seals and other areas of potential air leaks, cleaning out filters and traps, etc), or was tempted to cut corners with feedstock preparation. With a small downdraft gasifier, problems don't necessarily reveal themselves straight away, but they can lead to hours or days of downtime if the reactor has to be emptied. One hour spent preparing and being vigilant can save eight hours trying to repair the results of carelessness. Complacency comes into this as when a system works well, there is a tendency to ease back on diligence with operating and maintenance tasks (including feedstock care).

Operation and maintenance should not therefore be looked on as an additional burden. It is as important as switching the system on and connecting it to a power transmission system. Gasifiers don't have failsafe mechanisms and like filling the reactor with poor wood, the problem is only apparent once it is too late. Systems are not fool-proof, another reason why the self-sufficient person is the one who makes it work.

appendix C - diagnostics

Temperature and tar determination are the best ways to understand what is going on inside the gasifier thermodynamically. It the gas is too hot and the engine power output is poor, then possible reasons could be an air leak (affecting the reduction zone), bridging, or too coarse a feedstock. Too low a temperature is a symptom of blockages (either through small feedstock and fines or slag build-up) as not enough air can get in and/or wet fuel.

Thermocouples (temperature probes) are a simple and effective way to monitor that the reduction zone is thermally optimised, preferably if one is positioned just below the throat and the other just above the ash grate/just prior to where the gas leaves the reduction zone. The upper thermocouple will reveal whether the combustion zone is getting to the desired temperature (and also perhaps not getting too hot!). The other will monitor the reduction zone exit temperature. The ideal gas temperature coming out of the reactor should be about 350°C, but can range up to 600°C. If at 800°C or above, it is telling you that the reduction zone is not working efficiently and something abnormal is occurring in the reactor. This might be due to air leaks. There is also the possibility of too high a temperature coming from above the throat, e.g. spreading combustion zone.

Measuring the ratios in the product gas between $CO:CO_2$, and $H_2:H_2O$ is another way to determine whether the reduction zone is healthy. There are gas analysers that can do this, but expect to pay at least £8000, so the best way to assess whether the reduction zone is optimised is by having thermocouples, or at the least by observing the colour of the flare stack flame.

As the gasifier system operates under negative pressure, having some form of manometer helps. At its simplest it can show that there is initial suction and therefore no gross air leaks. An air leak will suck air in which can cause localised increases in temperature that can cause damage and will also alter the thermodynamics of the systems resulting in poor quality gas. A number of pressure sensors are even better as they can identify where localised pressure problems are occurring.

Table 2: causes of non-ideal reactor status that manifest themselves by poor gas quality.

Air leak	Air creates combustion. Too much will cause the combustion zone to spread into other zones. Localised combustion in the reduction zone means char is combusted rather than used for reduction. A high temperature effect.
Bridging	Void spaces are created inside the reactor which fill with O_2 and there is less endothermic pyrolysis reactions to cool the system. A high temperature effect.
Feedstock too coarse	Causes bridging and large void spaces which result in the problems as described above. A high temperature effect.
Fines	Dense packing. Restricts throughflow of gas (air) and radiative heat transfer. Has a low temperature effect.
Wet feedstock	Water imposes a high parasitic energy drain on the reactor, reducing internal temperature and therefore causing higher tar and lower quality gas. A low temperature effect.

Nozzle damage, or too wide nozzle distances	Cold spot in centre of reactor through which dirty (and colder) gases pass.
Reduction zone not functioning	Hot and smoky gas produced as not enough endothermic reduction reactions.

glossary

Activation Energy: an energy barrier which needs to be overcome for chemical reactions to occur. This is achieved through the input of energy such as bringing a lighted match into contact with paper.

Air: the oxygen content of air is 21% O_2. The remainder is 78% nitrogen, 0.9% argon, 0.03% CO_2 plus traces of other light molecules. It is often sited as evidence to support intelligent design that if the concentration of moleculer oxygen in the air were to vary slightly by just a few %, then it would mean the end of human life on earth.

Autothermal: a system (the gasifier) operating at steady state without any external input of energy. Some portion of the feedstock is sacrificed to create the energy to maintain the equilibrium.

Biomass: literally any organic matter of biological origin, but conventionally this refers to non-fossilised solid organic matter, such as wood, straw, and soft vegetation. Biomass is distinct from "Biofuels", this latter term defining crops which when harvested are used to create liquid chemicals such as bio-diesel or bioethanol, usually on a large agricultural scale.

Brash: leftover forestry clippings and trimmings such as leaves and thin stems.

Bridging: when the feedstock binds together and does not flow as desired through the system. A void space is created under the "bridge".

Calorific value: the energy available in a material if it were to be fully combusted. This can either be gross (also known as Higher Heating Value – HHV), or net (Lower Heating Value). Gross calorific value (HHV) assumes that any steam released in combustion would be cooled back to liquid form so that this latent heat of vaporisation is recovered. Net calorific value (LHV) excludes this additional energy.

Channelling: when the feedstock binds together at the sides of its container and only flows through the centre section. Often "channels" form from collapsed "bridges".

Combustion: a chemical reaction that occurs when a fuel is heated in the presence of oxygen such that the oxygen combines with some of the molecules in the fuel. Remember the fire triangle (fuel plus heat plus oxygen) results in a flame. Air contains 21% oxygen, so unless using pressurised cylinders in laboratory or industrial applications, all combustion occurs using the oxygen in air. Biomass and fossil fuels are combusted in an open fire or in a boiler chamber, or in the cylinder of an engine. Flaming combustion, as its name states, is when flames emanate from the fuel. Incineration (literally, 'burn to ashes') is another term used for combustion, primarily in the large scale waste disposal industry. The combustion reaction releases energy in the forms of heat and light.

Compound: relatively stable chemical substance formed from elements and molecules, e.g. CO_2 = carbon dioxide, H_2 = molecular hydrogen, H_2O = water, steam, or ice, and longer chain hydrocarbons.

Element: the smallest chemical unit that cannot be decomposed. In chemical notation, represented by a letter or combination of letters, e.g. C = carbon, H = hydrogen, Fe, = iron, etc. Elements contain an atomic nucleus and orbiting electrons. Elements combine with each other to form molecules.

Endothermic: of a chemical reaction. Absorbs heat from its surroundings.

Equivalence Ratio: the amount of oxygen used for gasification to achieve the highest purity producer gas. It is the ratio of this amount compared to that which would be required for total combustion of the feedstock.

Exothermic: of a chemical reaction. Liberates heat to its surroundings.

Feedstock: the term "feedstock" is used to describe the "fuel" of a gasifier. There is no literal demarcation between the use of the two nouns in this context, however "feedstock" refers to the reacting material, like the fuel of a car. But whereas the fuel of a car goes into the engine, with a gasifier, the feedstock creates a fuel *i.e.* a gaseous product for use in engines.

Fines: small pieces among the feedstock *i.e.* dust and other fragments below the size specifications required (usually ≤ 2 mm diameter). Fines are created by the wood chipping process, particularly with drum chippers or where the cutting blades are not sharp, and also by attrition during movement of the feedstock both inside and outside of the gasifier.

GC-MS: gas Chromatography-Mass Spectrometry. Analytical technique for identifying tar molecules.

Heating rate: how quickly a feedstock is exposed to maximum temperature. Gasifiers usually have a slow and gradual heating rate, which due to pyrolysis principles, favours gas production over liquid (tar) production.

Hydrocarbon: a molecule containing the elements carbon and oxygen, but usually also hydrogen.

kJ: a unit of energy (0.00028 kWh).

Manometer: simple device for gauging changes in pressure. Can be made from a piece of transparent tubing and some water.

Mole: a quantity of a chemical substance. Mol is short for mole. In the case of chemical reaction nomenclature (e.g. R1 to R9) it refers to all the molecules in the given reaction. The number of moles in a volume of gas is independent of the type of molecule, and molar volume of gas depends on temperature (decrease temperature and get more molecules of gas in a smaller volume) and pressure (increase pressure and get more molecules of gas in a smaller volume).

Molecule: combination of elements. The smallest part of a compound that can participate in chemical reactions.

Municipal Solid Waste (MSW): municipal solid refuse. Mixed waste from residential and commercial origin. Excludes sewage sludge and certain hazardous materials. Also usually refers to only the audited fraction of municipal waste that is collected.

NO$_x$: nitrogen oxides (N_2O, NO, and NO_2). Atmospheric pollutant that forms in combustion. Involved in ozone depletion, photochemical smog, acid rain, and global warming.

Phase: solid, gas, and liquid are chemical "phases". Substances can exist in different phases depending on the conditions of temperature and pressure. Therefore imposing a change in temperature (and/or pressure) can result in a substance changing phase from solid to gas to liquid or back again.

Polymers: complex molecues made of many individual elements. Usually these are hydrocarbons so contain variously bonded and arranged carbon, hydrogen and oxygen.

Pressure drop: the difference in pressure between two points. It reveals and quantifies the resistance to gas flow caused by friction or obstructions in the system.

Producer gas: see Syngas

Product: new substance(s) formed from a chemical reaction.

Pyrolysis: is defined as the unoxygenated heating of a substance. Pyrolysis forms part of the gasification process rather than being synonymous with it. A gasifier by its nature must "pyrolyse" the wood, but the gas which comes from simple pyrolysis is a dirty gas which would not be of a quality for downstream engine applications. Gasifiers are designed to have adaptations which will clean this gas, and this is the total aim of gasification design – to create the highest calorific value as possible with the least impurities.

Reactor: a chemical "reactor" is a vessel in which a chemical reaction occurs. Not just nuclear reactors, but also cooking pans on a stove are chemical reactors.

Reactant: the initial substance(s) in a chemical reaction.

Residence time: the time taken for the feedstock to pass through the gasifier, or specifically the duration of its exposure to controlled conditions of temperature and reaction chemistry.

Retort: a container in which the modern process of charcoal manufacture occurs by pyrolysis.

SO$_x$: sulphur oxides (SO$_2$ and SO$_3$). Atmospheric pollutant linked with respiratory problems and acid rain.

Stoichiometry/Stoichiometric: describes the balancing of chemical reactions, *e.g.* the quantity of each product and each reactant that is required for the complete reaction. For example, in the following chemical equation, one molecule of methane, reacts with two molecules of oxygen to form two molecules of water and one molecule of carbon dioxide. Note that whole number examples are used ubiquitously in text books, but reaction chemistry is far more complicated than this.

$$CH_4 + 2O_2 \rightarrow CO_2 + 2H_2O \qquad \Delta H = -890 \text{ KJ/mol} \qquad R6$$

Syngas/Producer Gas/Generator Gas: syngas (full title synthesis gas), is a gas mixture containing predominantly CO and H$_2$. The Swedes called gas from a gasifier "Generator Gas" during the Second World War (27). It was also called Suction Gas in the decades before this (27). The name "Producer Gas" is also used for gas from a gasifier. The origin is unclear, although one author suggests that it derives from the fact that no storage system is used and that the gas is supplied directly following production (25).

Thermocouple: temperature sensor. Low voltage metal probe that can tolerate temperatures of at least 1000°C.

Tuyere: air nozzle.

Upstream/downstream: upstream refers to before the region and downstream after. Imagine a water wheel accepting river flow. Water before it enters the wheel is "upstream" and after it has left the wheel is "downstream". Chemical processing also uses these terms as it deals with material (gas, solid and liquid) flows.

symbol nomenclature

ΔH: this is heat change at constant pressure. It is best thought of as the amount of energy that can be released or absorbed per unit of reacting substance. For reactions which give off heat, such as combustion, the value of ΔH is succeeded by a minus sign to show that energy is released to the surroundings and this type of reacton is termed "exothermc" (*exo-* = external, and *thermo-* = heat). ΔH can also be positive, and these reactions, termed "endothermic" (*en-* internal, and *thermo-* = heat) are ones which require energy input, for example boiling a pan of water...... or pyrolysis. Note that it is not uncommon to find mistakes with other books where the signs are reversed.

N: Normal Conditions. Because gas concentration varies as a function of temperature and pressure, the "N" is used to standardise it. Under "normal" conditions, pressure is taken as atmospheric, and temperature is either $0\,°C$ or $20\,°C$. $1\ Nm^3$ of gas weighs about 1 kg, so $1\ mg.m^{-3}$ approximates to 1 ppm, and 1000 ppm = 0.1%.

chemical species

C = carbon. Solid at Normal conditions.

CO_2 = carbon dioxide. Gaseous at Normal conditions.

CO = carbon monoxide. Gaseous at Normal conditions.

CH_4 = methane. Gaseous at Normal conditions.

H_2 = hydrogen. Bimolecular gas at Normal conditions.

H_2O = water or steam, or ice.

HCl = hydrochloric acid.

N_2 = nitrogen. Bimolecular gas at Normal conditions.

O_2 = oxygen. Bimolecular gas at Normal conditions.

S.I. numeric prefixes

T = tera = 10^{12}

G = giga = 10^9

M = mega = 10^6

k = kilo = 10^3

m = milli = 10^{-3}

μ = micro = 10^{-6}. This unit of length is also called a micron.

n = nano = 10^{-9}

chemical reactions

$C + O_2 \rightarrow CO_2$	ΔH = -394 KJ/mol	R1
$H_2 + \frac{1}{2}O_2 \rightarrow H_2O$	ΔH = -242 KJ/mol	R2
H_2O (liquid) $\rightarrow H_2O$ (gas)	ΔH = +41 KJ/mol	R3

$(C_6H_{12}O_6)_x \rightarrow$ heat without oxygen
$(H_2O + H_2 + CO + CH_4 + .. C_5H_{12})$ ΔH = + ?   R4

$CO + \frac{1}{2}O_2 \rightarrow CO_2$	ΔH = -111 KJ/mol	R5
$CH_4 + 2O_2 \rightarrow CO_2 + 2H_2O$	ΔH = -890 KJ/mol	R6

$C + CO_2 \rightarrow 2CO$ (Boudouard reaction)
ΔH = +172 KJ/mol   R7

$C + H_2O \rightarrow CO + H_2$ (water gas reaction)
ΔH = +131 KJ/mol   R8

$CO + H_2O \leftrightarrow CO_2 + H_2$ (water gas shift reaction)
ΔH = -41 KJ/mol   R9

equations

Cold gas efficiency =

$$\frac{\text{LHV of gas x gas volumetric flow rate}}{\text{LHV of feedstock x feedstock volumetric flow rate}}$$ (Eq. 1)

other lowimpact.org titles

See lowimpact.org/about/lowimpact-org-publications

 **The loveliest Loo:** tells of a girl's unexpected discovery of a different kind of toilet, a compost toilet... and one with a surprise! *The Loveliest Loo* will make you laugh as you ponder the most basic elements of life and how we regard them. A charming story about a beautifully simple way to conserve our natural resources. The striking black and white images are designed to be coloured using pencils.

Timber for Building: from growing and felling trees, to selecting the right wood for the tasks you have in mind, this book explains how to efficiently convert low value local round wood to high value sawn material and get the best out of your equipment, before outlining different drying methods and taking you on to preparing the timber for your project.

 **Food Smoking:** in our cave-dwelling days, food smoking was used to preserve food and then our ancestors discovered just how great it makes food taste. This book covers the basics of cold and hot smoking; delves into the principles of combustion and explains brining and dry salt curing, plus how to source wood for smoking and provides plans for building a cold smoker and smoke generators.

How to Build a Wind Pump: the wind pump described in this book can pump rainwater, greywater, river, pond or well water for irrigation, aerate a fish pond, run a water feature or even be a bird scarer. The turbine is 700mm diameter, and the head plus rotor weighs less than 4kg. In a light-to-moderate wind, it should pump 1000 litres per day, with a head of 3-5 metres.

Herbal Remedies: teaches you to identify, grow and harvest medicinal plants. It shows you how to make a range of simple medicines including ointments, salves, syrups, oils, compresses, infusions and decoctions. There are sections on body systems, explaining which herbs are useful for a range of ailments, and detailed herb monographs. This second edition has been revised to take account of recent changes in UK legislation.

Make your own essential oils: a fascinating hobby, or for the professional aromatherapist, a way of ensuring that your products are fresh, unadulterated and organic. This book also describes how to make creams, lotions, balms, gels, tinctures and other skin-care products from the essential oils and distillate waters you have produced.

Wind & Solar Electricity: there are chapters on the various system components required, how to put them all together, batteries, grid-connected systems, and there is even a basic electricity primer. Andy has analysed the output of his system for over 10 years, and these real-life figures are included. Developments in the associated technology and UK government incentives have led him to make substantial revisions and additions for this second edition.

Make Your Own Natural Soaps includes both hot and cold process soap making, with step-by-step instructions. There are extensive bar, liquid and cream soap recipes, full details of the equipment needed to make a start and a re-batching chapter just in case anything goes wrong! And for anyone interested in turning their new skills to profit there is information on the legislation and regulations you need to comply with to be able to sell soap.

Heating with Wood: covers everything you need to know about wood heating, from planning a system, choosing, sizing, installing & making a stove, obtaining & storing firewood, and cooking with wood, to heating your water with a back boiler. It includes chainsaw use, basic forestry, health & safety, chimneys, pellet and woodchip stoves, and how to light a fire and keep it going.

Compost Toilets : reduce water usage, prevent pollution and produce fertiliser. Built properly they can be attractive, family friendly and low maintenance. Contains everything you need to know about building a compost toilet, plus proprietary models, decomposition, pathogens and hygiene, use and maintenance, environmental benefits and troubleshooting.

Solar Hot Water: particularly applicable to domestic dwellings in the UK, although the principles described are widely adopted throughout the developed world. This book provides a comprehensive introduction to every aspect of solar hot water, including relevant equipment, components, system design and installation and even how to build your own panels.

How to Spin: a veritable encyclopedia of spinning know-how. Comprehensive instructions allow new spinners to get started with the minimum of equipment and give those who have a wheel already a full understanding of its operation. The chapter on 'other fibres' offers a wealth of information about fibres as diverse as yak and SeaCell, as well as information on the preparation and spinning of silk

notes

Lightning Source UK Ltd.
Milton Keynes UK
UKHW020842280822
407815UK00002BA/9